NAVAL AIR ENGINEER

A personal story of the Fleet Air Arm in wartime

By
Hugh R Langrishe
Lt (A/AE) RNVR

Published by:

HRL Aviation
7 Stonehill Close
Bromyard
HR7 4XD

Printed by Orphans Press Ltd
Leominster, Herefordshire
HR6 8JT

ISBN: 978-1-3999-1408-6

Cover photo: The author on home leave in Edinburgh in July 1943

Table of Contents

Hugh Langrishe (right) with his father, Lt Col. John du P. Langrishe DSO RAMC. and his mother, Helen Dorothy Langrishe, in Summer 1943

Foreword by Nick Blackman OBE

During the four years of its existence during the First World War the Royal Naval Air Service (RNAS) made spectacular advancements and achieved astonishing successes, yet politicians back in the UK believed improvements could be made by ordering the merger on April Fool's Day 1918 of the army's Royal Flying Corps (RFC) and the navy's RNAS to form a third service. The ensuing 19 years were a bleak period for naval aviation but in 1937 the Admiralty won back its Fleet Air Arm under full naval control and a race against time ensued to recover from the Royal Air Force's inter-war neglect to re-build an efficient fighting Service.

By late 1942, mid-way through the Second World War, the Fleet Air Arm was expanding rapidly with freshly trained aircrew and mechanics arriving and deliveries of both new aircraft and aircraft carriers becoming available. The Admiralty recognised that naval aviation was becoming more complex and that a new cadre of specialist technical officer was needed; the naval Air Engineer Officer (AEO). The navy recruited some from its own ranks of Engineer Officers manning the engine rooms of its surface warships, but it also turned to universities, aeronautical colleges and aircraft manufacturing firms.

Hugh Langrishe joined the ranks of this 'rare breed' direct from his undergraduate course at Edinburgh University, underwent naval training and found himself assigned to a naval air station in the UK. At the second Quebec Conference in Sep 1944, President Roosevelt accepted Churchill's offer to deploy the Royal Navy alongside the US Navy and US Marine Corps against the Japanese in the Pacific. As a result, the Royal Navy machine swung into action to send a force of some 260 ships and a quarter of a million men. This powerful carrier-based strike force needed a new type of aircraft support organisation ashore, both in Australia and in the allied-captured Pacific islands, to sustain its fighting edge. The sheer enthusiasm and competence of the Royal Navy's first generation of young AEOs seeking to achieve this quickly made a difference and the Admiralty recognised that appointing such officers to individual squadrons would be beneficial.

Hugh Langrishe was at the forefront of all these historic changes, deploying to Australia in support of the British Pacific Fleet (BPF) and himself becoming one of the first ever Squadron AEOs in the Fleet Air Arm. Quite apart from being a fascinating and humorous personal recollection of a young undergraduate going to war, this short memoir also constitutes a uniquely well-written and priceless historical perspective on the high-water mark of the Royal Navy's Fleet Air Arm as it became one of the most powerful naval air arms on the planet, second only to the US Navy.

Nick Blackman OBE

Captain Royal Navy (retired)

Oct 2021

Glossary of Terms

Note: Over 70 years, changes have taken place in the Royal Navy. It is important to understand that no efforts have been made to relate the descriptions of Naval ranks, rates, specialist designations, Naval Establishments etc. to those used in the twenty-first century. Everything quoted in this glossary of terms was in general use during wartime in the 1940s. As explained in the story itself, the RAF influence on the technical aspects of the Fleet Air Arm was at that time still strong, while many supervisory roles were still performed by flight sergeants and sergeants of the RAF until late in the war, and indeed afterwards.

(A)	Used with rank, Air Branch (FAA) Officer.
(A)	Used with AF or AM to denote "airframe", cf (E) for engine.
A25	Accident Report form partly completed by pilot with much dread.
A/AE	Air Branch, Air Engineer Officer, with rank
AA1,2,3 or4	Air Artificer 1st to 4th class; AA1 to AA3 rate as CPO (Chief Petty Officer), AA4 as PO (Petty Officer)
AAF	Auxiliary Air Force (not "Royal" until after the war)
ADC	Aide de Camp
ADDL	Aerodrome Dummy Deck Landings (practice carrier landings on airfield runways at Naval Air Stations)
ADO	Air Defence Officer, airfield defence
AE/P	Air Engineer Pilot Usually one on each station for test flying duties
AEO/AirEO	Air Engineer Officer, usually spoken of as an "airyo"
AF	Air Fitter, either (A) Airframe, (E) Engine, (L) Electrical or (O) Ordnance
AGO	Air Gunnery Officer
AGS	Aircraft General Spares
AID	Aeronautical Inspection Directorate
ALO	Aircraft Electrical Officer
AM	Air Mechanic, either (A) Airframe, (E) Engine, (L) Electrical or (O) Ordnance
AP	Air Publication. Body of Air Ministry publications covering everything needed to run an air force, admin to technical, flying and ground, manuals and parts lists for every aircraft and engine.
AR	Aircraft recognition
ARP	Air Raid Precautions
ARS	Aircraft Repair Section or Squadron. RAF term adopted by FAA
ASDO	Air Station Daily Order
BALBO	Large formation of aircraft, named after General Balbo (Italy) who led a formation of Italian Flying Boats from Italy to America and back in 1933.
BPF	British Pacific Fleet
CAG	Carrier Air Group. Group of squadrons embarked together.
CAP	Combat Air Patrol
CBW	Combat Bomb Wing, (USAAF)
Cdr	Commander RN. A senior officer
Cdr (E)	Officer in command of an air station technical department. Usually a Commander.
Cdr (F)	Commander Flying, at air stations or embarked, in charge of all flying. Known as "Wings"

Cdre	Commodore. OiC RN Barracks. Temporary rank for a substantive Captain in this appointment. At time of writing a "one-star" field officer, and now a recognised rank.
Chippy	Carpenter
CO or OiC	Commanding Officer or Officer in Command
Col	Colonel (Army or Royal Marine)
CPO	Chief Petty Officer. A senior rating
CVE	Aircraft Carrier (Escort)
Defaulter	Rating who has broken rules or regulations appearing before Captain or Commander for punishment.
DFC	Distinguished Flying Cross (RAF), awarded for gallantry in action.
DH	de Havilland Aircraft Company
DLCO	Deck landing control officer or "batsman"
DSC	Distinguished Service Cross (RN) Awarded to officers for gallantry in action. (DSM for ratings)
(E)	Engineer, marine. Used with rank and designated with a purple ring.
E/P	Engineer pilot. Variant of A/EP
ERS	Engine Repair Section or Squadron. RAF term adopted by FAA
FAA	Fleet Air Arm. The official use when Air Branch returned to Royal Navy in 1938/9 but in general use throughout the war.
FANY	First Aid Nursing Yeomanry. Ladies who work as ambulance drivers attached to army units.
FIDO	"Fog, intensive dispersal of". Installed at a number of RAF airfields to assist bombers returning home in foggy conditions. Burnt large quantities of aviation fuel alongside runways
"Flight"	Common name for Flight Sergeant
FRU	Fleet Requirements Unit. Flying unit for cooperation with Fleet Air Defence
F700	Serviceability form unique to each aircraft; signed each day when required to fly, by air mechanics and pilot
Gash	Rubbish
GES	General Engineering Shop/Section/Squadron. Adopted from RAF
Goofer	Off duty people who hang around to watch flying activities, particularly on a carrier. Also the open air places where these people gather to watch, mainly on the island.
GS	General service, i.e. not "hostilities only"
HMS	His Majesty's Ship. Applies to any ship in Naval Service, ashore or afloat
HO or h.o.	Hostilities only. Denotes type of engagement in service. Normally applies to all recruits in wartime.
KDG	Kings Dragoon Guards. Originally a cavalry regiment but mechanised in WW2
KR and AI	Kings Regulations and Admiralty Instructions. The way by which the Royal Navy is administered. (Now known as Queen's Regulations for the Royal Navy)
(L)	Electrical branch. Officers show green strip, quoted with rank.
LMSR	London Midland and Scottish Railway
LNER	London and North Eastern Railway
Lt Cdr	Lieutenant Commander, showing 2½ rings on his sleeve; ranks above Lieutenant and below Commander
Lt Cdr (F)	"Little F", Lt Cdr Flying, assistant to Cdr F. Normally a Lt Cdr (A) in wartime
Lt Col	Lieutenant Colonel in Army or Royal Marines. Ranks above Major and below Colonel. Equivalent to Cdr RN

Master at Arms	Senior rating in a ship. Reports to Commander on all aspects of discipline on the lower deck. Addressed as "Master" or "Sir" by lower decks or as "Master" by officers.
MATMU	Mobile Aircraft Torpedo Maintenance Unit, a subcomponent of a MONAB tasked with torpedo servicing and supply. These units were attached to MONABs supporting disembarked TBR (Torpedo, Bomber, and Reconnaissance) Squadrons. Torpedoes were never used by the BPF.
MM	Mobile Maintenance Units (1 etc), a technical component of a MONAB
MNAO	Mobile Naval Airfields Organisation, the Headquarters and parent unit for Mobile units whilst forming in the UK.
MONAB	Mobile Operational Naval Air Base – Sent to Australia and Far East in 1944 onwards as complete self-administered units. A total of 9 were deployed
MS	Mobile Servicing Unit, a component of a MONAB specialising in one type of aircraft.
MT	Motor Transport
MU	Maintenance Unit. Term used by RAF for all ancillary non-flying establishment, technical, stores etc. (cf FAA use of ARY, Aircraft Repair Yard)
NASTM	Naval Air Station Technical Memoranda
NavEx	Navigation exercise. Flight for specific purpose of practising navigation by the observer.
No.1	First Lieutenant in Naval Establishment or ship, used in common parlance.
O	Observer. Officer member of aircrew in multi seat aircraft, responsible for navigation and for "spotting" enemy ships and aircraft. Carries "O" wings on left sleeve above rank ring.
O	Designated ordnance rating, eg AF(O) or AM(O)
OC Troops	Serving officer in overall charge of all uniformed passengers in a troopship
OS2U	Observation Scout aircraft (2U) is the second aircraft in this class built by Vought, a manufacturer, later bought by Sikorsky. US Navy designation.
OTU	Operational Training Unit. The last stage in qualifying as operational aircrew
P&RTI	Physical and Recreational Training Instructor
Plumber	Colloquial name for Engineer Officer
PMO	Principal Medical Officer in a ship or shore establishment; not always but usually a Cdr, known as Surgeon Commander and carried red rings in his rank. In charge of all things medical and commonly known as "the Quack"
PO	Petty Officer. Non-commissioned officer who lives "below decks" Has supervisory function
POAM	Petty Officer Air Mechanic. In writing add speciality, (A), (E), etc
PRU	Photo Reconnaissance Unit
QARNNS	Queen Alexandra's Royal Naval Nursing Service
Queen Mary	Common name for a long low articulated road vehicle used for transporting aircraft, usually partly dismantled, or the remains of a crash.
RA	Royal Artillery
RAAF	Royal Australian Air Force
RAF	Royal Air Force
RAFVR	Royal Air Force Volunteer Reserve
RAMC	Royal Army Medical Corps
RAN	Royal Australian Navy

RANAS(N)	Rear Admiral Naval Air Stations (Northern)
RANVR	Royal Australian Navy Volunteer Reserve
RDU	Receipt and Despatch Unit. Shore establishment which collects and distributes complete aircraft, usually in conjunction with an erection unit. Also dismantled aircraft for spares. In Australia incorporated in a MONAB
RM	Royal Marines
RN	Royal Navy
RNARY	Royal Naval Aircraft Repair Yard (naval equivalent to RAF MU)
RNAS	Royal Naval Air Station
RNB	Royal Naval Barracks
RNH	Royal Naval Hospital
RNR	Royal Naval Reserve
RNVR	Royal Naval Volunteer Reserve
RNZAF	Royal New Zealand Air Force
RNZNVR	Royal New Zealand Naval Volunteer Reserve
RP	Rocket projectile
RSL	Returned Serviceman's League. The Australian equivalent of the Royal British Legion
RTO	Rail Transport Officer (provides rail tickets)
(S)	Supply and Secretarial branch; officers carry white band with their rank. Senior officer on naval air stations either a Commander or a Lieutenant Cdr. Colloquially known as the "Pusser". This branch also supplies the Captain's Secretary, usually a Lt (S). Senior officer referred to as "Pay"
S of AE	School of Aeronautical Engineering RAF Henlow, Courses for Junior Officers
S of TT	School of Technical Training, RAF designation for a school for teaching recruits the basics of their trade, even cooks, cleaners and clerks.
SAP	Semi armour piercing bullet or shell
Sick bay	Ship's hospital. Used colloquially to mean the medical department, fully equipped with life-saving apparatus, ashore as well as afloat.
Subbie	Normal usage for a Sub Lieutenant
TAG	Telegraphist Air Gunner, third member of TBR aircraft crew
TAMY	Transportable Aircraft Maintenance Yard
"Tilly"	Ford utility passenger vehicle based on a Ford 10 van, the "Utilecon".
TBR	Torpedo Bomber and Reconnaissance aircraft
TCD	Trinity College, Dublin (Dublin University)
U/S	Unserviceable; of aircraft, not to fly
USN	United States Navy
USS	United States Navy Ship
VAD	(Red Cross) Voluntary Aid Detachment. Volunteer Nurses not enlisted
Very Light	Signal pistol for firing multicoloured flares, invented by Samuel Very 1877
VIP	Very Important Person
WAAF	Women's Auxiliary Air Force
WAO	Warrant Air Officer
WO	Warrant Officer
WRNS	Women's Royal Naval Service (aka Wrens)

Introduction

This book describes my wartime service in the Fleet Air Arm (FAA) as an Air Engineer Officer. However, there are many people, even in the 21st century, who are unaware that the Fleet Air Arm is an integral part of the Royal Navy and is in no way a part of the Royal Air Force. All personnel in the Fleet Air Arm hold Naval ranks and rates. It was unfortunate that for a period of twenty years from 1st April 1918, when the RAF was formed from the Royal Flying Corps and the Royal Naval Air Service, until 1938/'39, everything connected with Naval aviation was manned, organised, managed, maintained and practised by the RAF. Almost all pilots were RAF officers but when on board an aircraft carrier they became equivalent RN officers and wore Naval uniforms. All maintenance personnel were RAF aircraftmen, working under Warrant Officers, Flight Sergeants and Sergeants of the RAF who had been trained in RAF Apprentice Schools or Schools of Technical Training. It was the RAF which adopted the name "Fleet Air Arm of the RAF". At the time of the creation of the RAF virtually the entire Royal Naval Air Service was transferred to the RAF, including many experienced officers who later rose to Air Marshall rank. Virtually no one with experience transferred to the Royal Navy so when the new Fleet Air Arm came into existence in 1938 it adopted RAF methods en bloc. Certain technical training schools were designated for Naval ratings and the resulting qualifications were navalised RAF ranks, e.g. air fitter and air mechanic (A) or (E) for "airframe" or "engine". Those men who passed out from the apprentice schools were "Air Artificers", at first in a grade 4, classified as Petty Officer, but rising through the grades to AA1. These held Chief Petty Officer rate, thus aligning with traditional Artificers in the ships' engine rooms. The nomenclature of aircraft workshops in the RAF was adopted, thus ARS (Aircraft Repair Section), ERS (Engine Repair Section) and GES (General Engineering Section). As the Service evolved additional titles came into use for instruments, radio, radar, armaments, torpedoes etc. and the training of such men and women was done in the RAF training schools. All Air Branch officers carried an "A" in the curl of their rank insignia; pilots and observers wore their wings above the curl on their left sleeve.

During those early days the Navy gradually increased the number of aircraft carriers in commission – HM ships *Argus*, *Hermes*, *Eagle*, *Furious*, *Courageous* and *Glorious* in the twenties, followed in 1937 by the first of the purpose-built big ships, *Ark Royal*. As landing techniques evolved so the numbers of aircraft embarked could be increased. Ships did not spend their entire lives at sea so, when they were refitting or re-storing in their home bases, the squadrons would fly ashore for training in all their skills, particularly keeping in constant practice for deck landing, a technique which was entirely different to landing on a runway. Whatever exercise pilots took off to do they would always land in deck landing mode – approach at a speed just above the stall with the engine on power and cut the throttle on the command of the deck landing control officer – the DLCO, or "batsman", whereupon the aircraft would drop on to the deck to pick up an arrester wire. There were no aerodromes (as they were called in early days) specifically allocated to the FAA. Flights and, later, squadrons disembarked to RAF stations which were convenient for ships docking – mainly Netheravon, Upavon in south Wiltshire, Gosport in Hampshire, Leuchars, Evanton (for Invergordon naval base) and Donibristle in Scotland.

Lee-on-Solent, Hampshire, came into wider use in 1934 when a grass airfield was laid down to augment its use as a seaplane base. FAA landing techniques were recognized by the provision on naval airfields of four hard runways of shorter length and of narrower width, than for RAF airfields. They were provided with plenty of Admiralty Type 'S' hangars so that aircraft could be stowed with their wings folded when not in use. Several airfields, Arbroath being one, had a dummy deck marked out on one of the runways.

No sooner had the Admiralty regained control of the FAA than a programme of constructing airfields exclusively for the Navy began. On completion they were commissioned as Royal Naval Air Stations (RNAS). Between 1939 and 1944 seventeen were built to Admiralty specification, between St. Merryn in Cornwall and Hatston in the Orkney Islands. A number of other airfields built for the RAF were passed or loaned to the FAA, including Fearn in Easter Ross, Maydown in Co. Londonderry and others. Culdrose and Brawdy were both built after the war. On the outbreak of war in 1939 (or perhaps before), RNAS Lee-on-Solent became the principal Naval airfield. It was to be the initial base for newly appointed Air Engineer Officers and those engineers awaiting appointment before or after courses. The FAA adopted a different approach for the designation of its second-line training squadrons from the RAF. Whereas the latter named the unit, for example No. 1 Service Training School, the FAA would number it as a squadron in a sequence in the 700 series. There were to be one hundred such squadrons.

Blackburn "Baffins" of 811 Squadron FAA, flying over their ship HMS Furious in 1935. The identity of each aircraft is shown by the numbers on the fuselage sides.

In the mid-thirties, with future expansion in mind, the Admiralty introduced a scheme of specialist aircrew. Trained specifically for the Navy, either as pilots or observers, young men were granted short service commissions and were identified by having an 'A' in the curl of their rank identification. Known as the "A Boys" they were a very important part of the FAA during the war. As many as fifty-six of these pilots fought in the Battle of Britain, either seconded to the RAF, or they were in Naval squadrons which took part in the battle. At the same time, it became obvious that there would be a shortage of pilots for the second-line squadrons. The Admiralty, with the help of the Royal Aero Club, circularised all holders of a civilian pilot's licence to invite them to join the Air Branch of the Royal Naval Volunteer Reserve for second-line duty. The invitation was successful and many with previous RAF experience, and some also with a technical background, joined the Service. They did a valuable job.

FAA front-line units, which had their own sequence of numbers in the RNAS, were for some fifteen years designated as Flights from 400 upwards. The whole numbering system was revised in 1933 when Naval squadrons were numbered in the 800 series. During WW2 additional numbers were allocated in the 1700s and 1800s. and eventually as many as 125 squadrons had been formed. A squadron would have anything between six and twenty-four aircraft.

And so, the Fleet Air Arm, with me waiting eagerly in its wings, went to war.

Hugh Langrishe

Bromyard, October 2021

PART ONE: BECOMING A NAVAL OFFICER

In the beginning

Background

My paternal grandfather Richard was a member of a well-established Anglo-Irish family. They had settled in southern Ireland at the end of the English Civil War in the 17th century and had acquired an estate centred on the village of Knocktopher, a few miles south of Kilkenny. Grandfather (b.1834) passed through the traditional route of fee-paying secondary school to doing articles with a civil engineer in Co. Galway, western Ireland. He did not go to university, as he had hoped, but became a civil engineer and architect through membership of the Institution of Civil Engineers and the Royal Institute of Architects of Ireland. His second son, John, my father (b.1883), went to St. Columba's College, Dublin. He had thought of becoming an engineer, but chose medicine and studied at Trinity College, Dublin (TCD) for five years. After graduating in 1906 he joined the Royal Army Medical Corps (RAMC) and served as an Army doctor for twenty years, retiring as a Lt Col. in 1928. He then started a second career in Edinburgh University as a lecturer in Public Health and Tropical Medicine (these two specialities were linked for many years) until his death in 1947. My mother, Helen, was the daughter of a London solicitor, Philip George Collins, and lived in Beckenham, Kent, at the time of her marriage in 1914. She lost her two elder brothers in the Great War. The second one, Lt Cdr Hugh Duppa Collins RN, Gunnery Officer in HMS *Monmouth*, went down with his ship in the Battle of Coronel in November 1914. His name is one of the first few on the Naval Memorial on Plymouth Hoe.

I had two siblings; my brother John was born in 1917, my sister Dorothy in 1921, and I was born in Bermuda in 1923 while my father was serving in the rank of Major in the garrison there as DADH (Deputy Assistant Director of Hygiene), a RAMC appointment. As one would have expected in the thirties, all three of us went through the private education system, almost entirely boarding. Both my brother and I went to Cheltenham from Aysgarth School, Yorkshire as candidates for an RAMC scholarship. John obtained a lower grade Exhibition, but I was unsuccessful at the second stage of the exams. John had already graduated from Cambridge by 1938 and had gained his Certificate 'A' in the Officers Training Corps (OTC) at school, and Dorothy and I were completing our secondary education as war approached. Personally, I was very much influenced by the national news of the time and the frequent reporting of the latest rearmament programmes for all three services. I experienced the very basics of military training in the school OTC, and by 1939 had also passed my Cert. 'A'. The national atmosphere at the time was one of rising concern at the imminence of war – with my fellow pupils we had dug air raid trenches away from our boarding houses in September 1938, only to fill them in again after Mr Chamberlain's visit to Germany to see Hitler and to report "peace in our time".

I had been interested in cars and aeroplanes since, probably, the age of six or seven. My brother, being six years older than me, was already a regular reader of *Autocar*, *Motor* and, later, *Motor Sport.* He became a solicitor in 1947. I can date my interest in aeroplanes to one day in 1929 or 1930 when my father, always interested in technical matters, took us to Turnhouse aerodrome, just outside Edinburgh, to look at the aeroplanes over the fence. It was the base of No. 603 Squadron, AAF[1] which had been formed in 1925 with the DH9A day bomber; it was re-equipped with the Westland Wapiti in 1930. Which type we saw I cannot remember, but I can remember that the aircraft were started by means of the Hucks Starter, a contraption used for many years on aircraft whose engines were too powerful to swing by hand or were out of reach. I started to collect postcard pictures of aeroplanes in around 1934 specialising in British and American aircraft. The collection now numbers over 4,500, and I have mainly used my collection to provide the illustrations in this book.

This interest developed through schooldays, past the stage of just wanting to be a pilot. I was a regular reader of the weekly magazines *The Aeroplane* and *Flight*. These stimulated my interest in the technical side of aeronautics, and I decided that I wanted to know how aeroplanes worked and were built. As with all public schools at that time the curriculum was based on the requirements of university entry – usually Oxbridge via the Oxford & Cambridge Universities Examination Board – which was traditionally classical and academic. The subjects included mathematics but no sciences. At that time, and until relatively recently, Cambridge University and TCD, and probably Oxford, required a School Certificate credit in Latin to gain entry, whatever subject you intended to read. Actually, I enjoyed Latin and obtained a credit. My prep. school curriculum was a part of this system, including Latin and, for scholarship boys, Greek, so I passed into that stream at Cheltenham – "Classical and Modern".

Airco Ltd. DH9A day bomber. Entering service in August 1918 on the Western Front, the "Ninak" remained in production and in the RAF and AAF until being superseded by such aircraft as the Wapiti between 1925 and 1931. It formed the initial equipment of many AAF squadrons from 1925 for up to 5 years.

I took my School Certificate in 1938 at fifteen and obtained four credits, including Latin, qualifying me to matriculate at Cambridge should I wish, my brother having gone up to Pembroke College in 1935. By that time, I had set my mind on aeronautical engineering. With some difficulty I had found that there were very few places offering aeronautical subjects and there was nobody at the college who could advise me, certainly not my housemaster. Two of the establishments I had found were the Chelsea College of Automobile and Aeronautical Engineering and the De Havilland Aeronautical Technical School, the latter based at the DH Aircraft Company at Hatfield. What I did find was that aeronautical engineering was rather looked down upon, and not only by universities; I believe that only Southampton University had initiated a course, probably under the influence of the local aircraft company Supermarine, part of Vickers-Armstrongs. The two technical schools mentioned combined

1 Auxiliary Air Force. Throughout I have either inserted a footnote with the meaning of any abbreviation on the first occasion of its use or explained the term on its first use. The Glossary at the start of the book includes all the abbreviations and explains any technical terms. I notice that the current online MOD "official" glossary of abbreviations runs to some 373 pages.

practical workshop training with classroom subjects which led to exams for professional institutions such as the Royal Aeronautical Society, the Institution of Production Engineers and the Society of Licensed Aircraft Engineers. Such qualifications opened the door to jobs all round the world, and not necessarily within the aircraft-building industry itself.

Westland Wapiti IIA J9835, 84 Sqn. RAF, Shaibah, Iraq, 1936. The scorpion badge was approved by King George VI as the new squadron badge in December 1936. These aircraft were classified as "General Purpose" and were used on policing duties in the 1920s and 1930s in the Middle East and the NW Frontier of India. This type equipped No. 603 Sqn., AAF from 1930 to 1934.

My housemaster was aware of my ambitions and had corresponded with my father. I cannot recall ever having had a serious conversation with my father but perhaps I did. On the evening I returned to school for the autumn term in 1938 my housemaster called me in to his study to discuss the next stage of my education. Out of the blue, and certainly without putting the choices to my father, he put them to me – I was only 15.8 years old: either I would go into the sixth form "Classical and Modern" and forget about engineering, or transfer to the fifth form in the "Military and Engineering" stream and take another School Certificate in 1939. This would be in a largely different range of subjects, dropping Latin but adding Physics and Chemistry, neither of which I had ever done. In fact, I had never been into a science laboratory in my life. Another was Geography, a subject I enjoyed but had dropped when I was eleven. Higher mathematics was another new subject. I was persuaded that this move would make an application for an engineering course at university more appropriate. I decided on the latter course and duly obtained six credits in the summer of 1939 – but not in Higher Maths, nor in Physics or Chemistry. In theory I still had two years' schooling in the sixth form to come, but I fitted none of the criteria for the existing sixth forms! I was placed in Lower 6th Science, to work on a syllabus which was as strange as a foreign language. It was not long before I came to the conclusion that this was a complete waste of time, so in the Christmas holidays I suggested to my father that he should advise the College bursar and my housemaster that I would leave at the end of the Easter term of 1940, and this I did.

In the meantime, my father had discussed my ambitions with university colleagues and with a friend who was a mining engineer. Not one had any knowledge, let alone experience, of the aviation world, its industry or its education system, and they all looked down on the idea! One gentleman (of course they were all men) even suggested that I should take a mechanical engineering degree and then go to the Parsons turbine works on the Tyne as a graduate premium apprentice. Through my father's position in the University he could claim special terms for an undergraduate placing, and this was arranged. But there was a snag: the Scottish examination system was quite different to its English counterpart and Edinburgh would not accept School Certificates from an English examination board unless one could produce an acceptance letter from an English university. So, my father went back to Pembroke College, Cambridge, submitting both my school certificates: they accepted the first one with its credit in Latin and rejected the second one with its six credits but no Latin! Thus was I accepted by Edinburgh University to read mechanical engineering. I would live at home. I started in October 1940. I was not unhappy, but I struggled with Chemistry and Physics and with advanced Mathematics while finding the engineering subjects enjoyable and interesting. The system of mathematics tutorials, with as many as twenty students at a time with one lecturer, just did not work with me and became a complete waste of time, making matters worse. End of year exams in 1941 were a disaster, as were "re-sits". But, surprisingly, the war came to my aid.

Miles Magister N3780 basic dual-control trainer. Shared all initial training in the United Kingdom with the DH Tiger Moth. Issued to some RAF Stations as the Station Flight communications aircraft but at first also used for flight experience for University Air Squadron cadets. RAF Turnhouse provided a Magister for the writer's first flight, piloted by the UAS CO, Sqn. Ldr. H.M.:Pinfold. N3780 served in No. 15 Elementary Flying Training School at Carlisle and with No. 21 EFTS at Booker

By late 1939 the country was engaged in a massive re-armament programme and manning of all units was done by voluntary recruitment. My brother John had volunteered in September 1939 and he was called up a month or so later for officer training at the Royal Artillery OCTU (Officer Cadet Training Unit) at Catterick Camp in Yorkshire. Subsequently he was "trooped" out to Egypt to join a regiment of the Regular Army artillery, the 7th Medium Regiment, in whose 25/26 Battery (The "Battleaxe" Battery; it was a surprise to learn that the origin of the 25th Battery was in Kilkenny in 1801.) he was to serve throughout the war except for a period of some twenty months when he was a POW in Italy. At the Armistice in September 1943 the commandant of the POW Camp opened the wire to enable the prisoners to escape into the countryside. For his epic walk along the Italian mountains back to the Allied Lines John was Mentioned in Dispatches. Spurning an offer of staff work in London, he rejoined his regiment at Felixstowe early in 1944, received refresher training and went over to France on D+3 with his battery. He was in northern Europe until the Armistice on 7th May 1945, frequently in action.

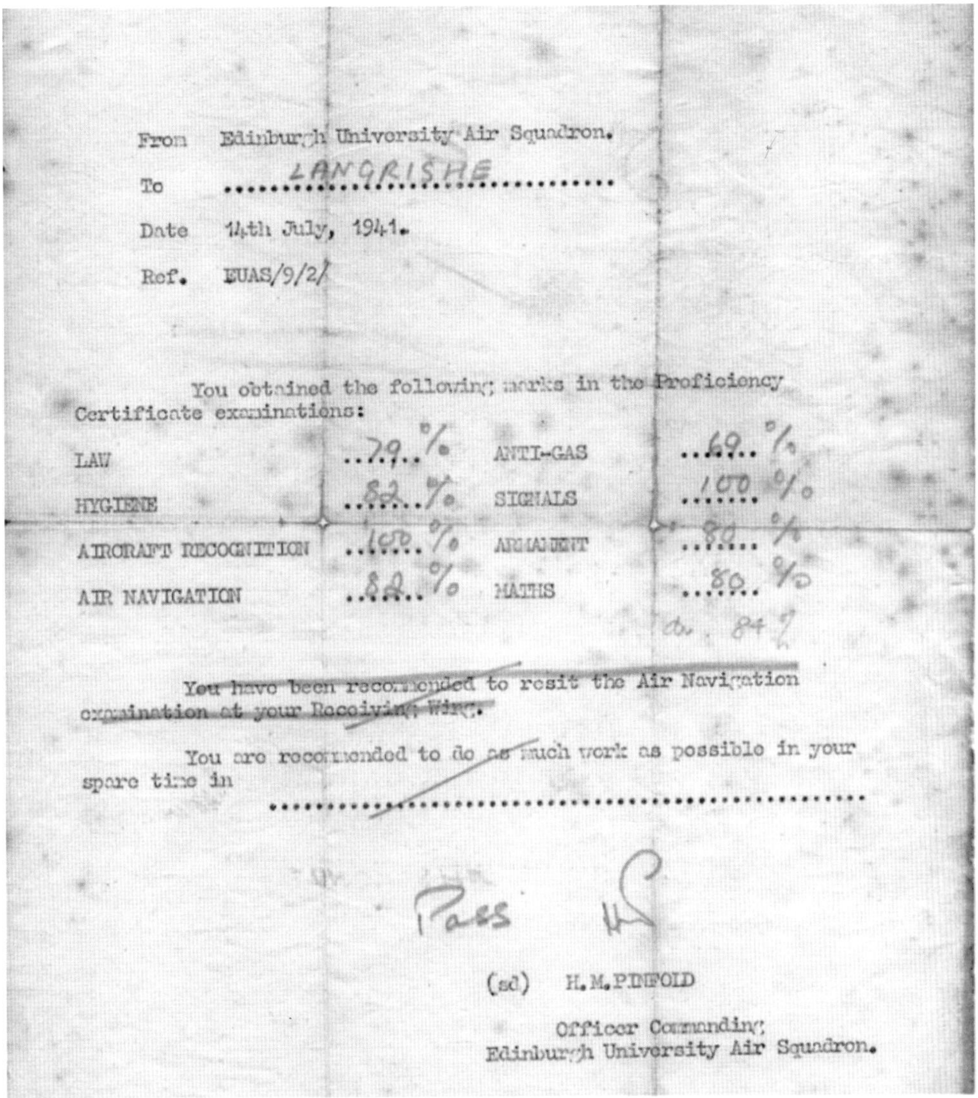

From Edinburgh University Air Squadron.

To ....LANGRISHE..........

Date 14th July, 1941.

Ref. EUAS/9/2/

You obtained the following marks in the Proficiency Certificate examinations:

LAW	..79..%	ANTI-GAS	..69..%
HYGIENE	..82..%	SIGNALS	..100..%
AIRCRAFT RECOGNITION	..100..%	ARMAMENT	..80..%
AIR NAVIGATION	..82..%	MATHS	..80..%
			av. 84%

~~You have been recommended to resit the Air Navigation examination at your Receiving Wing.~~

You are recommended to do as much work as possible in your spare time in ....

Pass

(sd) H.M.PINFOLD

Officer Commanding
Edinburgh University Air Squadron.

Edinburgh University Air Squadron: Course No. 1 Pass Certificate, 14th July 1941

Following the trend to "do her bit", my sister Dorothy joined the Red Cross VAD, the Voluntary Aid Detachment which had first formed in the Great War to provide auxiliary nurses to serve in the field hospitals. After preliminary hospital training in an Emergency Hospital in Fife she was transferred to the RN Hospital, Kingseat, Aberdeenshire, an establishment which had been the County Mental Hospital in peacetime. Here she became a Royal Navy VAD. After a period of general nursing she joined the X-ray department and qualified as a radiographer. She was then posted to the Emergency Naval Hospital which had been set up in the hotel on the Gairloch, a short distance from Loch Ewe which was the assembly point for the Arctic convoys to Murmansk. There she attended to the men who had endured terrible hardships in the Arctic. With all three children in uniform my father undertook to open a Civil Defence First Aid Post in an empty house opposite to where we lived. He planned the equipment, found volunteers to man the post, trained them and organised their routine. All this was done in addition to his daily university lecturing.

Edinburgh University Air Squadron, No.1 Course, March to July 1941. In centre, seated, is Sqn. Ldr. H.M.Pinfold RAF, CO, on his right is F/Lt. Mackay, Adjutant and Staff Officer. Cadet H.Langrishe is seated, fourth from right, Cadet Ron Flockhart is in the second row, second from right. Ron became a successful racing driver in the fifties, driving the V16 BRM and then the Ecurie Ecosse D-type Jaguars, winning the 24hr race at Le Mans in 1956 and 1957.

This leads me back to my own experiences in the university. The university had an Army Senior Training Corps – no other Service at the time had a similar cadet force – and I joined. But early in 1941 the Air Ministry formed training squadrons at a number of provincial universities, Edinburgh being one, and the Edinburgh University Air Squadron came into existence. (see Appendix A). The squadrons were effectively units of the Air Training Corps, i.e. students were not enrolled into the Service, with a small staff seconded from the RAF, e.g. the Officer Commanding was Sqn Ldr H. M. Pinfold RAF, a career officer and fighter pilot who had been in action during 1940. Subject only to interview, and without a medical examination, students were signed on. The object of the course, which was part-time, was to go through the equivalent syllabus and subjects of an Initial Training Wing, i.e. the ground school which preceded initial flying training. There was no flying instruction, but all students had an air experience flight at Turnhouse in the station flight Miles Magister, piloted by our Commanding Officer. Later, squadrons were issued with a Tiger Moth, housed and serviced at the nearest RAF airfield. There were forty-three students in the first course. In the final exam in July 1941 I passed with an average of 84%, including 100% for aircraft recognition and for "signals" (morse code). I had proved to myself that I could pass exams!

Later in the year, while awaiting call-up, I helped instruct aircraft recognition to the second course. The possibility of students wanting to resign from their courses and join the armed services had been recognised by the Government with the formation of Joint Recruiting Boards at all universities, the object being to identify potential officers. This was the next step I took. The Boards comprised active or retired officers of all three services. My father was appointed Chairman of the Edinburgh Board and I knew three or four other members of the board, including the Air Squadron CO. At the

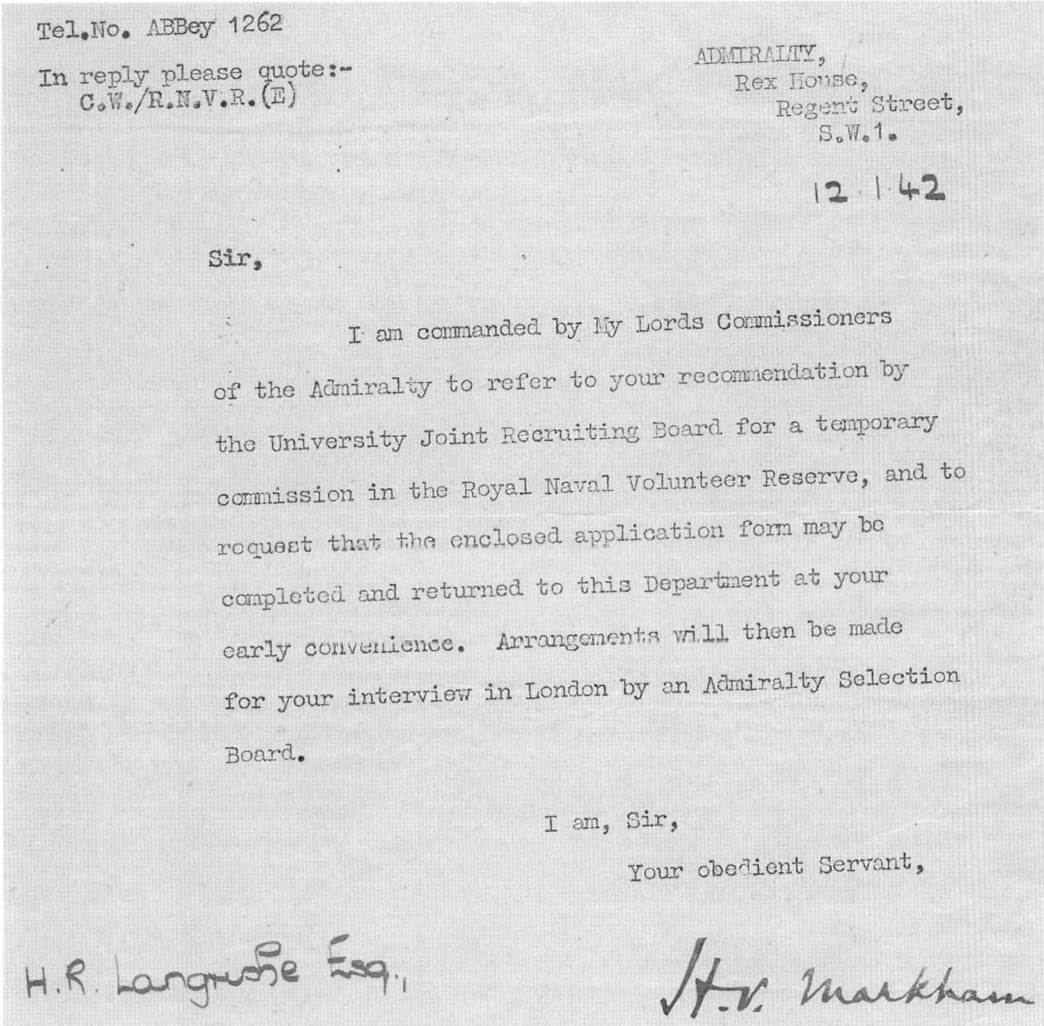

Tel.No. ABBey 1262

In reply please quote:-
C.W./R.N.V.R.(E)

ADMIRALTY,
Rex House,
Regent Street,
S.W.1.

12.1.42

Sir,

I am commanded by My Lords Commissioners of the Admiralty to refer to your recommendation by the University Joint Recruiting Board for a temporary commission in the Royal Naval Volunteer Reserve, and to request that the enclosed application form may be completed and returned to this Department at your early convenience. Arrangements will then be made for your interview in London by an Admiralty Selection Board.

I am, Sir,
Your obedient Servant,

H.R. Langrishe Esq.

H.V. Markham

Official letter of acceptance to serve in the Royal Naval Volunteer Reserve, subject to interview. The signature of the writer, Mr Markham, a senior civil servant had taken up his post as permanent secretary tot he Admiralty in 1941.

time of presenting myself to the Board I had failed the aircrew medical which the whole course had undergone. This was due to short sight for which I had worn glasses since the age of ten. So I knew I would have to impress the Board if I was to be recommended for a commission in another branch of the RAF; I had not, at this time, given a thought to the Fleet Air Arm. However, the Board obviously thought I was officer material and in the autumn of 1941 my name was chucked into the pool managed by the Ministry of Labour, responsible for allocating men and women to the appropriate service. It was a surprise when I was informed that I was to join the Fleet Air Arm (FAA) and not the Royal Air Force. So, the die was cast and within a few days, in January 1942, I was summoned to London for interview and medical at what I subsequently learnt were Queen Anne's Mansions. The doctor who examined me, a Surgeon Lt Cdr RN, had done his medicine at Edinburgh University and had experienced my father's lectures. I surmise that the interview was conducted by an RN gentleman whose main aim was to determine if I might become a "good officer". Anyway, I passed both stages, my short sight confirmed. Not long afterwards, and rather to my amazement, I was informed that I had been appointed to the Royal Navy Volunteer Reserve (RNVR) Air Branch as a specialist officer, i.e. an Air Engineer Officer. I was to be a Probationary Temporary Acting Sub-Lieutenant (A) RNVR, with pay of 9/- per day -- I was nineteen years old. I had no technical qualification, but I had boundless enthusiasm and many years' accumulated knowledge of aeroplanes and Army and Air Force cadet training. While awaiting call-up I made a point of keeping fit by spending three or four mornings a week in the University gym, enjoying one-to-one instruction from the staff.

Thus in March 1942, at the age of nineteen and two months, I proceeded to Royal Naval Barracks (RNB) Portsmouth, *HMS Victory*, to undertake what was known as a Divisional Course of a fortnight's duration. The course had been set up as an introduction to the Royal Navy for new Reserve officers of all branches – the various sleeve colours were bewildering. There were no more than a handful of officers destined for the Fleet Air Arm (FAA), not all of them air engineers.

My sister, Dorothy Langrishe, in her uniform as a Royal Navy VAD. Pictured during Christmas 1942

The early days

As was probably the case with most of the officers on this course, I was borne on the books of another establishment but detached "for course". My base attachment for the first sixteen months of my service was to be the Royal Naval Air Station, Lee-on-Solent, which by that time had become the main establishment of the Fleet Air Arm.

Because of all the Naval activity at Portsmouth Naval Base and Dockyard, the RNB Wardroom was bulging. Officers on short courses were put out to "B & B digs" and I was allocated to a terrace house in Southsea. I shared a room with another "new" Air Engineer Officer (AEO), S/Lt (A/AE) Ridge RNVR. This was my first encounter with an example of the wide variety of candidates for our branch. He was twenty-nine years old; at first, he had served as a merchant seaman but at the time of his call-up he was an air fitter (E) at Croydon! How on earth did that happen? We took all meals in the RNB Wardroom Mess, a room of impressive size and beauty. Due partly to the use of this mess for training officers' stewards, the walls were lined with young men eager to take your order. Unconscious gestures frequently resulted in two stewards taking the same order, causing embarrassment to both parties; none of us had experienced such personal service in our lives.

The course assembled in St. John's School, close to the barracks, under its CO and chief instructor, Lt Cdr House RN. The curriculum was ship-based and traditional. We paraded for Divisions every day, usually to the rousing accompaniment of the band of the Royal Marines. This often led into an extended period of parade-ground drill – something which I had been doing since I was eight years old and which I enjoyed. In the present context I felt it was overdone. Subjects of lectures and demonstrations were as follows:

- Parade-ground drill (lots of it)
- Unarmed combat
- Daily divisions and prayers
- V & W class destroyers (the latest at the time); description and organisation
- Shooting on .22 range; scored 34/40
- Air Raid Precautions (ARP), particularly in relation to RNB. (we were regularly detailed for ARP duties)
- Naval Officers; duties
- Royal Navy ship types
- "Leadership"
- Naval organisation and King's Regulations and Admiralty Instructions (KR & AI)
- Duties of a cypher officer, a Naval Chaplain and a Divisional Officer
- Battle damage control
- Naval customs and discipline, welfare

- Routine for joining ship; guidance on lifesaving after the order "Abandon ship"
- Fleet Air Arm (the FAA officer expected could not come and was replaced by a general service (GS) member of the course staff – not encouraging)
- Anti-gas, including the practice gas chamber
- Naval light weapons
- Naval dockyard organisation
- Lifebelt testing and usage (in swimming pool), also swimming
- Drafting, pay rates and rules
- Address by Commodore RNB
- Aircraft recognition
- Anti-invasion measures and methods, including practical demonstration of hand-held explosives

It can be seen that no time was spent on the history of the Royal Navy, and certainly nothing on what had happened during the first sixteen months of the war. The escape of the German warships up the channel, which had taken place only a few weeks before, was looked upon as a destroyer and light coastal forces' action, with scant mention of the attack by 825 Squadron's Swordfish from Manston. But I have no doubt that many of those on the course who found themselves in traditional Naval establishments or ships during the next four years, with not a Fleet Air Arm aeroplane in sight, had been well served. At least we had all been well fed in the Royal Navy's principal wardroom.

As our small group left Portsmouth by truck for Lee-on-Solent on Monday morning, 23rd March 1942, there was excitement in the air – but not for me! I felt really ill. So, after completing my joining routine, I made for the sick bay. The MO sent me straight to bed where I stayed until the Wednesday evening, at which point it was obvious that I was seriously ill. The doctor thereupon summoned an ambulance and on the evening of 25th March I was whisked down the road to Royal Naval Hospital Haslar. I was diagnosed with left lobar pneumonia and put on the Serious Case list. So much for all the physical exercise I had put myself through at the university gym.

Thus began my service in the Fleet Air Arm!

Air Engineering Training

This illness was a disastrous interruption to my training. Pneumonia was at that time a very serious illness. I was put on to a new drug known as M & B which was reserved for Service patients. Having not received a letter from me for a fortnight, my parents were surprised to receive a telegram informing them of my admission to Haslar, and my illness. They were, naturally, very concerned, so my father telephoned Haslar and, being a Military doctor, was able to talk to one of the Naval Surgeons on an equal basis. This satisfied him that the treatment was of the best. Certainly, I was satisfied; I had a room to myself and the nursing and the food were both excellent. After fifteen days I was told that I was to be moved away from Haslar for recovery and recuperation and I was taken by ambulance to Park Prewett hospital, near Basingstoke, in peacetime the Hampshire County Mental Hospital. Several of the wings had been taken over by the Emergency Medical Service for use as a Service hospital. The officers' ward was on the first floor and was very light and airy. I was told I was anaemic, so I had a bed beside a window, enabling me to observe and "spot" all the aviation activity; Basingstoke was within the flight pattern of such aerodromes as Boscombe Down, Eastleigh, Worthy Down, Netheravon, Thruxton and Middle Wallop. There was hardly a day in my entire stay when poor weather stopped flying. The number of aircraft types I logged was seldom below fifteen; one day I saw twenty-one different types.

Shortly after my arrival I was visited by a Mrs Scott. It turned out that her husband was Col. Scott, RAMC (Retd.), an old Army friend of my father's, who was Superintendent of the civilian part of Park Prewett hospital and lived in a very nice house within the grounds. Mrs Scott and my mother had in the past also been good friends. My father had written to his old friend to tell him about me, and when I was on my feet I would walk round to their house for tea and civilised chat. The weather was so good that frequently I was able to stroll through the hospital grounds and I sat or lay down at the edge of the very fine cricket ground – watching aeroplanes, for which the Scotts lent me a pair of field glasses.

After twenty-four days at Park Prewett I said good-bye to the Scotts, thanking them for their generous hospitality, and proceeded to Lee-on-Solent to see the MO, re-assess the contents of my suitcase and obtain authority to go on sick leave. Partly in order to save the long journey to Edinburgh and partly to take advantage of better weather prospects in the south, my parents had arranged for me to go to the Isle of Wight to stay with my Aunt "Bun" Collins and her daughter Jocelyn at Bembridge; my aunt was the widow of my uncle Lt Cdr Hugh Collins RN, the one who had lost his life at Coronel in 1914. I spent a blissful month on the island, some of the time being entertained by my aunt's friends in the village. Through contacts I was able to spend a day on a test programme in a brand-new Fairmile D-Class Motor Gun Boat (MGB) no. 610. It had been built at a small waterside yard at Bembridge owned by Mr. Alex Miller and this was the first time it had been on the water. These boats were of approximately 90 tons; they were 115 ft. long with a crew up to thirty men. Propulsion was by four Packard-built Rolls-Royce Merlin marine-engines, each driving its own propeller. Together they produced 4,800hp and were expected to propel the boat at over 30 knots. Armament was as many as twelve quick-firing guns of various calibre, although for our test none of these were fitted. In their role of attacking German convoys close to the French coast the boats were found to be under-armed and many were fitted with two torpedo tubes, one on each side of the upperworks. They were re-designated as Motor Torpedo Boat. Our outing included a visit to Portsmouth to fill up with 2,500 gallons of 100-octane aviation fuel before we started tests in the Solent. Inevitably, as it was the first time the engines had run at sea, there were two faults to be rectified: one engine was losing oil and the thrust bearing of another was running hot. To everybody's disappointment these faults prevented us from undergoing a high-speed run on the test mile.

We had experienced a German air raid the first night I was at Bembridge. There was no other major enemy activity immediately overhead for the rest of my stay, although a raid on Cowes severely damaged one of the Saunders-Roe flying-boat sheds. On 1st June I reported back at Lee, hoping to return to duty at once. Actually, I found I had to go back to Park Prewett to undergo X-rays and to be cleared by the doctor there, taking a further two days. On return to Lee I expected the doctor to pass me immediately but the following day when I reported to Commander (E), and he in turn sent me to see Paymaster Lieutenant Jones, Captain's Secretary[2] - who was responsible for communicating courses to trainee officers - he referred me back to the Principal Medical Officer (PMO), who said he was awaiting my papers from Park Prewett! It took another two days before I was at last clear and on 6th June I was available for duty. It seemed that Lt Jones had no authority for appointment to courses; he was really only an agent for the relevant Air Branch officer, in London, to whom he had to refer. There could not have been any headquarters facility for Air Engineers at Lee – in fact, there was nobody delegated to welcome new officers to the FAA and to tell us what was likely to be ahead of us. Note that I had recorded the Lieutenant as "Paymaster Lieutenant". At that time rank nomenclature was still partly in the Victorian age. I recall that amongst the senior officers, the description of the head of the Supply & Secretariat Branch on a Station – those who had a white band on their sleeves – would be known as "Pay".

I had been off duty for two and a half months. I was now "awaiting appointment", in a category for which there was no specific work at Lee. The introduction of new entries of young officers to Lee-on-Solent seemed to be quite random; there was no "Introduction to the Fleet Air Arm", in a similar manner to the Divisional Course at Portsmouth, so we had absolutely no idea of the scope of the Branch, either ashore or afloat. For now, it did not seem to matter that I was totally inexperienced

2 In civilian terms the Captain's Secretary would be his personal assistant. Traditionally he was in the rank of Lt. (S), i.e. of the Supply and Secretariat branch but with special training for this role. He was the custodian of all the confidential records of the ship's officers, including detachment for courses, and dealt with all confidential signals from the Admiralty and Admiralty Fleet Orders and Confidential AFOs. He was in touch with the Air Engineer if technical matters were involved. By 1943 this post was increasingly being entrusted to WRNS officers of Second Officer or Third Officer rank who had specific training.

and had virtually no detailed technical knowledge of aeroplanes. The Aviation Electronics Technician, an officer whose identity escapes me, sent me to see WO Bagnall, the officer in charge of the main workshop (ARS) on the Station to give me something to do. Mr. Bagnall showed me round the hangar and gave me a Swordfish to supervise. But in the middle of the afternoon I was summoned to the telephone and told to get packed and off to Bristol to attend a course on the Taurus engine which had started that day. Once more I had to circulate to all the departments and obtain clearance to leave the Station, and to obtain a rail warrant. At this time officers in HM Services still enjoyed First Class railway travel so there was some comfort to offset the slow progress of the wartime cross-country trains. In the morning of Tuesday, 9th June I set off from Fareham, the train taking the route through a rather bomb-damaged Southampton to Bath and then to Bristol.

The Engine Division of the Bristol Aeroplane Co. had set up their school for Service personnel in premises at Henleaze, about a mile and a half from the Bristol factories. The place was a boarding educational school in peacetime and was fully equipped with single bedrooms and full catering facilities. We were made very comfortable. The Company had set up classrooms, a workshop and an exhibition room with many parts and sub-assemblies on display. This was a two-week course for Naval officers and senior rates. We were in the school all the first week, using all the facilities, which I found extremely interesting. The practical work included dismantling and rebuilding a Taurus engine and doing the same with a Claudel Hobson carburettor. We were free at the weekend to do what we liked, and I went up to Cheltenham to visit my old school. In the second week we were introduced to the Engine Division factory at Patchway, a factory built for Bristol under the Government's "Shadow Factory" scheme just before the war. We passed the whole day there, visiting the machine-shop, gear-grinding shop, engine assembly and, finally, in the aircraft factory itself not far away, engine installation into airframes – Hercules engines into Beaufighters. I was enthralled the whole time. It was an eye-opener to see flow-line production at its best. On the aerodrome we saw how these engines were air-tested experimentally using Fairey Battles. Another afternoon we visited the Colston engine assembly plant to see how engines were assembled on lines – components at one end, complete engines at the other, ready for test. Our third and final visit was to the factory down beside the docks – probably an old warehouse because it was on as many as three floors – where engines were stripped and rebuilt. It was surprising to see that even engines which had been taken from crashed aircraft and were covered or stuffed with clods of earth, were carefully dismantled and the parts scrutinised and tested so that they could be re-used. Rebuilt engines were only used for replacement of time-expired engines; new aircraft always had new engines. I thought that this course set an example for all manufacturers. In spite of the state of my technical knowledge, at no time was I out of my depth. I can attribute this to my years of close interest in aeroplanes and their engines.

It was on 18th June that I heard from home that they had had information from the War Office that my brother had been reported "missing in action" in North Africa. Now the waiting began for firm news of his fate.

The course ended at midday on Friday, 19th June. I had already been told that I was to report to Rolls-Royce aero-engine works at Derby the following Monday, for a Merlin course. I was allowed to remain at Henleaze that night, travelling on to Derby on the Saturday.

Bristol Beaufighter IF single-seater night fighter, the first aircraft specifically produced for this important task. Powered by two Bristol Hercules sleeve-valve engines, it had the wings, landing gear and tail unit of its predecessor, the Bristol Beaufort torpedo-bomber. The Beaufighter entered service in the summer of 1940.

There were three of us going on to this next course. One was Ross White, a charming man who had joined the Air Branch from the De Havilland Aeronautical Technical School, Hatfield. We met again in 1946 at Hatfield and regularly played Rugby together for the works' Rugby Club. The other man was Robbie Robertson, who I had first met at Lee-on-Solent and with whom I struck up a friendship which lasted for the rest of the war. As he was married it is likely that he had spent the weekend with his wife.

We had to find our own accommodation at Derby, although once in the engine school on the Monday we did get help from the chief instructor. Ross White had gone home to Harpenden, but he had asked me to find a bed for him as well on the Sunday. I spent Saturday night in the Midland – a Midland Railway hotel – but at 14/- for B & B I considered this expensive so moved out and into a small temperance hotel for the Sunday night. In my notes of these times I frequently refer to letter-writing. As a result of eight years of boarding school I was quite used to writing to my parents once a week, and they expected it to continue. Long-distance telephoning was not looked upon as normal but was for emergency use. In wartime it was expensive and, because of the special cheaper evening charges at times when men and women were off duty, it was renowned for long delays – even two hours was not uncommon.

When we arrived at the Rolls-Royce works in Nightingale Road on the Monday, we were surprised to learn that the course lasted only a week, whereas we had been under the impression that it was twice as long. I did not write down the number of men on the course, nor their ranks nor where they had come from. The instruction was in the hands of what I called "university men", I suppose to distinguish them from the "shop floor" technicians who had instructed us at Bristol. But whoever they were, all the instructors had been well selected. The Rolls-Royce people were very competent in both the theoretical and practical aspects of their course. They were aided by many detailed drawings and "cutaway" components. However, it was apparent that there was insufficient scope in the limited time available to examine the engine in very great detail. Taking out the time we spent walking round the works, we spent only four days in the classroom, of which about half was spent listening to lectures. These covered the engine as a whole, the SU carburettor and its fuel system, and the cooling and lubrication systems. For practical exercises we dismantled and rebuilt a port cylinder block and practised carburettor needle setting. I found the whole course fascinating. The first visit to the factory floor included the parts machine shop, forge, heat treatment department, the erection shop and the test beds. There were many marks of Merlin going through, Marks XX, 45 Series, 60 and 61 particularly, and we saw parts for the new Griffon in the machine shop. In a second factory visit we saw, in a rather dark and poky shop, where crankshafts and crankcase bearings were lapped and reamed. We also passed through the tool room where special grinding wheels were being made on sophisticated machines; in peacetime this had been the building to which customers had brought their Rolls-Royce cars for servicing. Rolls-Royce had always tested their aero-engines in open test beds so that visit was rather noisy. It was said that local residents were used to the noise and if it stopped it made them anxious in case something was wrong. When there was an air raid alert they were also subjected to clouds of smoke from the smoke generators which had been placed in all the streets surrounding the factory. The wide range of engines to be seen, which also included a number of Vultures, was explained to us as being because the Derby works did all the experimental engine work and manufactured small runs whereas the great Ford factory at Trafford Park, Manchester, made only Mk. XX series engines, for the Lancaster, Halifax, Mosquito, Hurricane, Beaufighter and Defiant.

Fairey Battle K9331, test-bed for the Bristol Taurus II sleeve-valve engine. Delivered to Bristol in 1938, it was taken out of service and scrapped in August 1942 after 571 flying hours

I noted in my diary that the senior instructor had directed us to a pub in the vicinity to satisfy our need for an alternative to the teetotal Carlton Hotel. We spent five very happy nights there, enjoying what was said to be the longest bar in Derby. Our midday meals were taken in a restaurant in a Company social building where, on one day, we mingled with a group of RAF pilots and ground engineers who included two Poles and an American from one of the Eagle squadrons. They were attending a course on engine handling.

It was obvious that the Derby works were not of 1930s construction, as had been the Bristol Patchway engine factory. The famous Nightingale Road works go back to early days of the Company. It was there that all the famous aero-engines and car chassis of the twenties and thirties had been designed and built. During wartime the roof lights had all been painted over so the workshops were all artificially lit, making them extremely gloomy. At that time the factory was working at least two shifts per twenty four hours, sometimes three. There was of course no sign of cars at the Derby works in 1942. All design and manufacture of car engines and chassis had been banished to an old foundry at Belper and temporarily suspended. However, the staff were put to developing an unsupercharged Merlin for use in Army tanks. Almost literally horrified at the specification of current tanks, Rolls-Royce, Belper, designed the tanks to accommodate their new engine. The engine was the Meteor and the first tank was the Cromwell, followed by the Challenger. Both were highly successful in action later in the War and the Challenger was still in service many years later.

Returning to Lee on Saturday, 27th June was quite an adventure. I asked the Rail Transport Officer at Derby station if I could get anywhere near Gosport without going through London. He routed me via Mangotsfield Junction, near Bristol on to the Somerset & Dorset Joint Railway. This took me via Bath and Shepton Mallet to Templecombe Junction. Here I had a pub supper before catching a train to Salisbury and down to Fareham. The weather had been wonderful, as befitted a long June day. It was an interesting journey, no doubt, but leaving Derby just after 11.00 a.m. and finally arriving at Lee at 10.30 p.m., it had been a long day.

Vought Kingfisher I (US Navy OS2U-1) FN656. Although delivered to the Royal Navy for seaplane tasks, this aircraft was allocated to the experimental establishment at Boscombe Down for trials as a landplane. Many were used in this form for communications duties.

Evidently I was told about my future because I noted that I would be under Mr. Bagnall in "workshops for the week that I am here". The structure of the Swordfish had until 1943 been covered entirely with fabric and sprayed with "dope". On the introduction of facilities for the carriage of rocket projectiles the underside of the lower wings was clad in aluminium sheet. The Swordfish biplane was by that time the only fabric-covered first-line spotter/anti-submarine aircraft in active service. There was variety: several OS2U Kingfishers were being converted from landplane wheeled undercarriages to a floatplane configuration of one central float and two smaller wing-mounted floats. They would be launched into the Solent down the original slipway dating back to 1917. OS2U Kingfisher was the standard two-seater spotter-reconnaissance floatplane carried by US Navy heavy ships and launched by catapult.

One aircraft was scheduled to go aboard the Armed Merchant Cruiser (AMC) *HMS Ranpura*, which had been requisitioned by the Admiralty in 1939 from the Peninsular & Oriental Steam Navigation Company for conversion to an Armed Merchant Cruiser with a main armament of eight 6" guns. These vessels overcame the drastic shortage of convoy escort ships. Her principal use, and of many other ships, was as convoy escort and she served in the North and South Atlantic Oceans and in the Indian Ocean between 1940 and 1943. By 1943 the need was lessening, and *HMS Ranpura* was bought by the Admiralty and converted to a fleet depot ship, serving until 1961 when she was broken up. The use of floatplanes on large warships of the Royal Navy was common in the 20s, 30s and early 40s, but by 1943 the concept of integrated Task Groups in major operations and the introduction of Escort Aircraft Carriers for convoys had made catapult seaplanes redundant and catapults were removed from all warships. The remaining seaplanes were converted back to landplanes by changing the undercarriage to wheels.

I hoped I might arrange a flight off the Solent in the Kingfisher but a choppy sea put paid to that – as it also did to a planned flight in a Walrus off the water. There was always plenty to do around the Station in the evenings. The weather was good and several of us went bathing off the nearby beach. There were informal dances to go to, with plenty of agreeable company from the Station's Wrens (WRNS). I was no expert at ballroom dancing, but it was fun – and good exercise.

Supermarine Walrus K8552, WP of 715 Ship's Flight, Far East, 1938, launching from HMS Suffolk. The Walrus was used throughout WW2 for Air-Sea rescue, from both coastal airfields and from aircraft carriers as well as from battleships and cruisers, until the latter ships lost their catapults during the war.

Three memories of Lee remain: the random placing of Dutch barns around the airfield, each containing one Swordfish with its wings folded; seeing a squadron of six Swordfish departing with bicycles lashed to the lower centre-section struts, and finally, amongst the Air Engineering sub-lieutenants roaming around the Station was one named Romanov. Yes, he was a member of the Russian Tsar's family. He had been educated in England and had attended the Chelsea College of Automobile and Aeronautical Engineering. He had a very friendly and suave personality. I never saw him again but I did see his signature on an aircraft document many months later.

I must have been informed, with a number of other sub-lieutenants at Lee at the same time, that I was to go to RAF Henlow for an extensive course; I just noted on 6th July that I had received my travel warrant for the following day. I learnt that had I not been away due to sickness I should have gone to Henlow for the course starting in May.

An officer I had met in the wardroom at Lee had given me an introduction to the editorial staff of *The Aeroplane*, at that time a weekly magazine produced under the banner of Temple Press. The editor was Peter G. Masefield and his deputy C. F. Andrews, both well-known aviation journalists. Andrews was also responsible for *Aeroplane Spotter*, a wartime spin-off from its parent magazine. I had at one time thought I might like to train as a journalist myself. So, on the morning of 7th July Robbie Robertson and I set off for London, allowing ourselves time for a visit or two before going on. I took the opportunity to call on my uncle who was the solicitor in Bedford Row, my mother's younger brother who had survived service in the Great War and with whose firm my brother John had started his articles in 1938. While I was in his office a letter arrived via the Red Cross from John giving the bare details of his present location as a POW in Italy. It was the first communication from him that we had had since his capture. We heard later that his battery had been surrounded by a German 88mm gun battery inland of Gazala in Libya on 1st June. John's battery had run out of ammunition and had no alternative but to surrender. After that unexpected piece of good news I went to the offices of *The Aeroplane* and made myself known to the editorial staff. I was shown round the premises, including the photograph and art departments, which were extremely interesting. After this a group of us met at Kings Cross and set off for Hitchin and Henlow. The latter was on a branch line from Hitchin and its station was in the heart of the camp.

Fairey Swordfish II in overall North Atlantic white finish. It is fitted with rocket projectile racks under the lower wings. It is here shown awaiting delivery from the Blackburn company's factory at Sherburn-in-Elmet. In the Fleet Air Arm these aircraft were often known as "Blackfish". The repainted Swordfish at Lee-on-Solent looked like this.

It is appropriate to comment on Lee-on-Solent as I found it in 1942. It was the principal air base of the Fleet Air Arm. Having started out as a seaplane training school in 1917, it did not get its grass aerodrome until 1934. Although the size was adequate for the time, its final asphalt three-runway system provided a longest runway of only 1,500 yards. This work was carried out as late as the summer of 1942. The only other Naval Air Station possessing a slipway for launching seaplanes was Hatston, Kirkwall, which opened in 1939. The transition of everything to the Royal Navy in 1939 must have been a nightmare. So few Naval people had had anything to do with aeroplanes or the men who operated or maintained them. There certainly had been little time to examine systems which had been standard RAF procedures for nearly twenty years. As far as I could see there were no obvious Fleet Air Arm staff supervising the induction and training of the young officers who were now joining the Air Engineering branch. It seemed to me that management of aircraft maintenance and repair needed a new approach. It was all too casual; although the Station buildings had suffered in an air raid on Portsmouth, the atmosphere was more of peace than war. I did not know it then, but my link to RNAS Lee-on-Solent broke on 7th July. I visited the place only once more: some sixty years later - the Fleet Air Arm Officers' Association enjoyed a commemoration lunch in one of the hangars, followed by a fly-past of historic aircraft and a ceremonial Sunset and lowering of the flag with the Band of the Royal Marines.

Air Engineering Training, RAF Henlow, No 16 (War) Course

I already knew that Henlow had a long history in the flying services. Its development, since its creation in 1918 as an Aircraft Repair Depot, had made it one of the largest technical units in the Royal Air Force. The sheds occupied a triangular area at the southern side of a splendid grass airfield. To cope with increased aircraft weights and busier traffic from the large No. 13 Maintenance Unit (MU) which had developed, some Sommerfeld steel rod and wire mesh tracking was laid. The grass grew through the mesh and the airfield retained its immaculate green. Considerable modernisation of the facilities had taken place in the thirties, including a new officers' mess and married quarters. Workshop sheds had been modified to accommodate the classrooms, lecture rooms and training workshops of the RAF School of Aeronautical Engineering, the school for officers which had been opened in September 1938. Air engineering at that time had no structure in the RAF requiring an officers' cadre. The experienced engineers of the 30s were the men who had gone through the two Apprentice Schools, Cranwell and Halton, and were Warrant Officers or senior NCOs. All commissioned officers involved with aeroplanes had been ex-Cranwell cadets, i.e. they were first and foremost pilots. It had been the custom in squadrons to appoint one of these officers as Squadron Air Engineer, through whom the senior NCO had access to the squadron commander. MU management structure worked in a similar way in that the various workshops were categorised as squadrons, flights etc, with an administrative squadron-leader in command. It seemed to me that all technical matters were in the hands of the highly motivated and very intelligent senior ex-apprentices, many of whom were warrant officers. It was a surprise to find that a substantial number of the skilled men on the shop floor in the MU were retired skilled RAF tradesmen who had been retained in their trades as civilians. Many of the School instructors were of the same origin; some of them had first served in the Royal Flying Corps. One of them, Mr. Borritt, had joined the RFC in 1913 and saw service with No. 3 Sqn, one of the first RFC units to go to France in August 1914.

The Camp, as it was known locally, was split in two, north and south of the A6001. To the north were all technical buildings, hangars and sheds, bounding the south side of the airfield, and to the south of the public road were the administration buildings, barracks, married quarters, officers' accommodation and sports and recreational facilities. The latter were magnificent: there were rugger, football and hockey pitches, tennis and squash courts, two good gyms, one of which was new, and in which a badminton court had been permanently marked, and a good station cinema. I was surprised that there was no officially organised PT or sporting activity for the officers and there was no incentive for the young officers to organise things themselves. That did not suit me, and I soon got involved with hockey – often "mixed" with players from the Women's Auxiliary Air Force (WAAF). Hardly a day of the summer and autumn passed without my participation in some open-air sporting activity. Early on I made arrangements to run a PT class for my fellow officers in the new gym, using the regime from the University gym classes we had while in the Air Squadron. Nobody in the School had ever done such a thing before. I had the use of the gym for an hour at least once a week between tea and supper and by the autumn I could expect sixteen men regularly, sometimes as many as twenty. I even attracted several members of No. 17 course which had started in September. So far as I am aware, there was no recognition of my initiative by the School staff. The Officers' Mess was hopelessly overcrowded. Meals were taken on a cafeteria basis, except for the Christmas lunch, and all officers "on course" were accommodated in the separate houses built in the thirties as married quarters, adjacent to the Mess. In most cases the two downstairs reception rooms housed two men each, the two main bedrooms the same, and a small bedroom took a single man. We were all Naval officers in our house, the single room being occupied by Lt Kitson. It was only a fifty-yard walk to the Mess. It was really a very satisfactory arrangement, and suitably quiet for study.

The School was a self-administered entity and occupied its own dedicated premises in the technical side of the Camp. The first thing we learnt was that the Captain's Secretary at Lee had contrived to get us to Henlow a day too soon. The course was to start on 9th July, not 8th! So, 16 War Course assembled at the school on the morning of 9th July. We numbered twenty-four, twelve Naval and twelve RAF officers. All the latter except one were Pilot Officers or Flying Officers of the RAF Volunteer Reserve (RAFVR), the odd one out being a Polish officer of equivalent rank. The Naval contingent comprised Lt Cdr (E) Smarden RN, Lt (E) Kitson RN and ten Sub-Lieutenants (A) RNVR. Both the RN officers were undergoing conversion to Air Engineering. Whereas almost all the Air Force officers were university graduates, the Naval officers were predominantly technical school graduates. Robbie Robertson was an exception. He was in his late twenties and was a Cambridge engineering graduate. It seemed as if I was the only one without letters after his name; at times I felt that I was very lucky to have been selected for this commission – and I was grateful that the Ministry of Labour had directed me to the Navy rather than the RAF. There were always three courses going through the School at the same time. 14 and 15 War Courses were ahead of us and when 14 passed out, 17 War Course entered. I cannot remember if No. 14 Course had been there during our first term, but it must have been. 15 Course included a number of Sub Lieutenants (A) RNVR with whom I became friends. 15 Course was replaced by No. 18 Course. The duration of each course was to be six months, broken into three terms with two weekends' leave. The turnover of courses took place at the end of each two-month term.

No 16 (War) Air Engineering Course RAF School of Aeronautical Engineering, RAF Henlow 30th July 1942 - From left to right, Top row, (in dark RN uniform): S/Lt(A) R.Robertson RNVR, S/Lt(A) H.Exley RNVR, S/Lt(A) K.Tryhorn RNVR, Middle row: S/Lt(A) Saville RNVR, S/Lt(A) Seaman RNVR, S/Lt(A) P.Cox RNVR, S/Lt(A) H.Langrishe RNVR Seated: S/Lt(A) Jones RNVR, Lt Cdr(E) Smarden RN, Lt(E) Kitson, RN, S/Lt(A) Sherwood RNVR S/Lt(A) G.Ainsworth RNVR

On that first morning we were issued with overalls, a toolbox and a set of textbooks, the details of the course were explained, and numerous forms were filled in. We underwent an intelligence test – said to be similar to those for seventeen year old boys, for which I obtained a grade which was "above average". The course comprised four main strands: theory, for example aerodynamics; practical, including use of hand and machine tools, casting and forging, structural repairs; dismantling, inspection and rebuilding of engines; laboratory testing of materials. Theory was taught in the classroom by a group of university dons who had been commissioned into the RAF for the duration of the war and who wore the uniform of Squadron Leader. Other lectures and demonstrations in the classroom, for example about aircraft systems or components, were usually delivered by a Flight Sergeant or Sergeant, and guest speakers visited regularly to talk about current affairs or their involvement in the aircraft industry. The mornings were normally taken up with practical work in the School's workshops and the classroom work was after lunch. We were divided up into groups of six or eight for the practical work and rotated through the various workshops in turn; the instructors were all civilians but they were all retired RAF tradesmen who had served for many years in uniform, some of them back into the Great War. An important part of the course was to be visits to the workshops of the MU. The various departments, all known as squadrons, covered the whole range of aircraft and engine manufacture. There were overhaul lines for Hudsons and Hurricanes, and Typhoons - shortly to be started, and an assembly line for Canadian-built Hurricanes which had been shipped in boxes in a stripped condition. The engine squadron overhauled Pegasus XVIII radial engines and tested them on open-air test beds. Another squadron overhauled DH-Hamilton and Rotol variable-pitch propellers, and another, gun turrets. Where necessary to complete an overhaul, parts were manufactured from raw materials, for example radiators for the Merlin engine. An instrument repair section had special dust-free workshops. All these shops were laid out to provide a flow of work from one end of the shed to the other, the final job being either a test flight or a bench test.

Lockheed Hudson V, QX-G of 224 Sqn., RAF. This aircraft was a conversion of the Lockheed 14 airliner first bought from the USA in military form in 1939 for Coastal Command. 224 Sqn. was the first front-line squadron to use the Hudson, the Mk.V coming into service in 1941. 13MU had an overhaul line for the Hudson.

Canadian-built Hawker Hurricane X AG162, EA-W of 55 OTU, RAF Unsworth, Tyne-and-Wear, in 1942. These aircraft were shipped from Canada in wooden packing cases in a partially dismantled condition. A production line was set up in 13MU at Henlow to unbox, re-assemble and test fly them.

On 26th August we were treated to our first view of a Hawker Typhoon single-seater fighter. This was the new Hawker fighter just entering service, an example of the new generation of aeroplanes designed around large engines, mirrored in the US by such as the Republic P-47 Thunderbolt for the US Army Air Force and the Vought F4U Corsair and Grumman F6F Hellcat for the US Navy and for the Royal Navy. These aircraft were all considerably larger than the familiar Spitfire and Hurricane, and the new Mustang from America, which, with a Rolls-Royce Merlin 60-series engine, was to become the allied air forces' greatest fighter. It was also a visitor to Henlow. We did not know it at the time, but the Typhoon and its Napier Sabre 24-cylinder sleeve-valve engine were both a source of great trouble and the cause of many lost lives. That day the Typhoon was only visiting, but within a few months a repair line was started at Henlow and the song of the Sabre rang out regularly over the aerodrome. During all this time I found myself acting as a sort of signal-box amongst my fellow students. In connection with sound recognition I seemed to be blessed with very acute hearing and whenever there was an unfamiliar-sounding aeroplane in the circuit I would know about it and we would escape from the workshops and make our way the several hundred yards to the aerodrome.

Also using the airfield were the two Whitley IIs of the Parachute Test Flight, a unit of 13 Maintenance Unit (MU) RAF. During the period of our course their principal task was the development of special parachutes for the Parachute Regiment, both for the men and their supporting equipment. Dropping was carried out on the airfield, and as it was at low level the Whitleys would drone around at virtually nought feet, rather incongruous amongst the more modern output of 13 MU. Actually, the Henlow unit had a proud history. It had been formed in 1925 and its aircraft had taken part in the annual RAF display at Hendon for many years, demonstrating the use of a parachute at low level. Platforms were attached to one of the interplane struts of the wings on each side of a Vickers Virginia bomber, the current equipment of the Flight. The aircraft took off with a man standing on each of these platforms; he was to release his parachute and be pulled off by the slipstream, landing a few seconds later on the airfield. It was a performance much appreciated by the spectators! We understood that it was only in 1941 that the last of the Virginias finally came to grief. It force-landed away from the airfield as a result of engine failure and was left to rot where it fell – or so the story goes. We were given the opportunity of a flight in one of these Whitley dinosaurs and were taken up three at a time, standing huddled around the pilot. On our flight there was an undercarriage malfunction which, fortunately for us, was overcome.

Hawker Typhoon IB fighter R8224. This aircraft was from the second batch of 250 aircraft, of which the first fifteen were built by Hawkers at Langley. The remainder were built at Brockworth by Glosters in 1942. For an early part of that year a white nose and black and white stripes under the wings were applied to enable Allied fighters and the anti-aircraft guns to differentiate Typhoons from the German FW190. A number of Typhoons were shot down by friendly fire. This aircraft was issued first to No. 56 Sqn. and then to No. 609 Sqn. It was struck off charge in 1943 after being overstressed doing aerobatics near Lympne.

The Vickers Virginia was the standard RAF heavy bomber between 1924 and 1937. A number were later used for training and experimental work; the Mark X equipped the parachute test flight at Henlow until the last one was written off in a crash in the vicinity of the aerodrome – in about 1941.

For the first term of two months the practical work involved the use of hand and machine tools, the latter being lathes and milling machines. We moved on to work on airframes, aero-engines and auxiliary equipment such as hydraulic and compressed-air systems, finally removing and re-fitting aircraft engines and making them run. We also spent time on test equipment to illustrate the theory we had learnt in the classroom. On most days about an hour was set aside for theory lectures and we had regular lectures by men from industry, one being Frank Lloyd, a member of the design team at Hawkers. We were expected to study in our spare time. Each student was expected to give a short lecture on some subject of his choice; mine was on Aircraft Recognition and it seemed to be well received. We worked a 5½ day week, usually finishing in time for tea in the Mess. There was time for sports between tea and supper and during the summer and autumn months I could use this time to participate in any one of tennis, squash, hockey and my PT classes. In the winter there would be rugger and badminton, and hockey if the ground was reasonably dry and smooth (there were no artificial pitches in those days). The Camp was large enough for there to be teams from the various squadrons, flights and departments or from, say, the WAAF or the Camp Hospital, to enable leagues to be set up; or trophies to be competed for. I joined in any game that was offered and took a part in organising teams from the School, either Navy or mixed, against all-comers. I taught several of my friends the basics of squash and hockey.

In wartime much store was set by organising social events, particularly to give both men and women the chance to socialise with the other sex. In such a big place as Henlow there were evening dances every week in the new gym under different organisers, and great fun they were. It was repeatedly made clear to us that we should not fraternise with the WAAF Airwomen, but what can you do to satisfy the yearnings of full-blooded young men? Perhaps the Naval contingent thought themselves out of the clutches of RAF retribution because a number of liaisons were formed; I found an attractive young woman who was as keen on games and music as I was and we spent many happy hours together, on and off the games field or off the Station attending concerts. There was a good Sunday train service from Henlow or Hitchin to London and good connections to the camp by bus or taxi. Groups of us would go up regularly, have a good meal, go to an orchestral concert in the afternoon, have another 5/- meal and get the last train back to Henlow. With no private cars we had to fall back on the trusty bicycle for trips outside the camp. We could get to Letchworth for a bit of shopping and during the summer evenings we bicycled to a flooded clay pit to swim, and just around the country for fun.

The Sopwith Camel was probably as well remembered after the Great War as was the Spitfire twenty-five years later. It was a single-seat fighter and was used in large numbers. It continued the use of the air-cooled rotary engine which was still in service in the 1920s in the Sopwith Snipe. The aircraft depicted is B3812. Front-line service was from mid-1917 for only a year, both by the RNAS and the RFC.

I found the practical work most interesting. I was pleased to find that I had an aptitude and an eye for precision work. As I had always thought, practising with the real thing made theory easier to understand, and I proved this to myself over these intensive months. At times I thought our practical instructors were too easily diverted from instructing to talk about their RAF service, but this did create "atmosphere". I was particularly intrigued by the stories of life in France in the Great War, tending to the likes of the FE2b, the BE2c, the Camel and the SE5, aircraft of legend. I do think that Service technical schools can waste time on out-of-date techniques; for example we spent many hours mending parts of the structure of a Hawker Fury fuselage. With the exception of the rear fuselage of the Hurricane fighter, such steel tube construction was not used on current Service aircraft. There would be enough skilled men already in service to look after the few of these aircraft, and their sisters the Hart family, in the training establishments which still used them. There happened to be a serviceable Hind single-engined light bomber at Henlow, which was flown one day by the Station CO with much flair.

SE5A Note the wartime censor has tried to eradicate the aircraft number but has allowed the more important squadron marking to remain. The SE5A line-up shown is of 32 Squadron, RFC, on a muddy "flying field" somewhere in northern France. The first squadron to serve in France was No. 56 Sqn., one of whose flight commanders was Albert Ball, a recognised "ace" pilot. The aircraft saw the war out. This was the first aircraft to fit the Vee-8 water-cooled Hispano-Suiza engine, later to be produced by the Wolseley company in England as the Python and Viper."

I struggled with some of the theory, particularly metallurgy, and I think this was because it was theory; I had found it difficult as a first-year subject at University, but after seeing, using and treating all sorts of metals during the war, I chose metallurgy as an optional subject in my Production Engineering exams in 1949 – and passed well. To start with I found the lack of Service formality in the management of the Course a little disconcerting. One would have expected to have periodical formal parades but this was not the case. There was no orderly procession from the mess to the School, we just made our own way in our own time. Someone in authority must have had a complaint or suggestion that this was not following standard Service procedure because late in the year the rules were changed; we had to assemble outside the Mess in the morning and after lunch and march down to the School. Some of the Course were rather put out by this, mainly I think because they were unfamiliar with parade-ground drill and found marching in step quite an ordeal.

16 War Course drew to a close just after Christmas 1942. The seasonal activity generated by Christmas rather interrupted the usual end-of-term celebrations. We gave a party for the teaching staff in the Red Hart in Hitchin a week before Christmas. I noted that while almost all of the workshop instructors were present, only one of the group of academic lecturers turned up. We were not impressed. We don't appear to have had any formal meeting with, or address from, the School CO or senior staff, which strikes me as unusual. We dispersed on Wednesday, 30th December, to leave "awaiting appointment". Little did I know that I was to return.

Air Engineering Training, RNAS St. Merryn: The Qualifying Course

I took my leave in the south, staying with my Collins solicitor uncle in Kent. It was while there that I received notice to report to RNAS St. Merryn, Cornwall, on 14th January 1943. The previous day, armed with my travel warrant, I went to Waterloo and found that I had a day's journey ahead of me to faraway Cornwall. Familiar faces appeared either at Waterloo or at stops en-route and it was obvious that all our Henlow crowd were destined for the same place. Our train carried the name boards of what was (still) the "Atlantic Coast Express" which, after over three years of war, showed a degree of optimism by the Southern Railway. Departing at 10.59, we followed the old Southern route to the south-west via Andover, Salisbury, Templecombe, Sidmouth Junction to Exeter, thence to Okehampton where the Plymouth coaches were detached. Then followed a stiff climb around the north side of Bodmin Moor on what was now a single track. It was wild country, accentuated by the howl of the westerly blowing through the undercarriage of the train. We arrived at Padstow at 18.20. The truck from St. Merryn was found to be unwilling or unable to manage the men and luggage it was confronted with, so it had to make two journeys to and from the Air Station. Thus, on a dark and windy midwinter evening we were finally deposited at the wardroom at St. Merryn. I don't suppose any of us had any idea of what we might find at this remote Air Station, or indeed even where it was.

Armstrong Whitworth Whitley GR VII Z6633 of No 612 Sqn Coastal Command, a similar aircraft operated by the 10OTU detachment from Abingdon at St.Eval.

The topography of the north Cornish coast is tortuous, to say the least. When we stepped out of the wardroom on our first morning, we could not even see the airfield! The officers' accommodation had been built on the north-east side of the airfield where the ground drops away from the plateau on which it is laid out. Extensive flat places are rare in north-west Cornwall. Three of the string of five airfields down the coast, of which St. Merryn is the northernmost, were built in most peculiar shapes in order to fit in three runways. The surrounding land was either wet or very wet. These five airfields, either on the cliff tops or on their own high points between the deep coombs, St. Merryn, St. Eval, St. Mawgan (Trebelzue at first, now Newquay Airport), Perranporth and Portreath were all very busy. Fighter, bomber and reconnaissance tasks were all in evidence, as were such second-line tasks as target-towing and anti-aircraft gunnery co-operation. As well as its Coastal Command operational function, St. Eval was used for final operational training by a detached flight from 10 Operational Training Unit at RAF Abingdon, using white Whitley VIIs. From time to time one would disappear out over the Atlantic, most often due to the navigator having steered the pilot on a reciprocal course further and further from land. St. Eval was also used as a United States Army Air Force (USAAF) staging post on the route to North Africa. After the long runway was opened at St. Mawgan in 1943, this airfield became a terminal for transatlantic flights as well as Prestwick. A FIDO installation was completed at St. Eval at the end of 1943. FIDO stands for "Fog, intensive dispersal of". Installed at a number of RAF airfields they burnt large quantities of aviation fuel alongside runways to assist bombers returning home in foggy conditions.

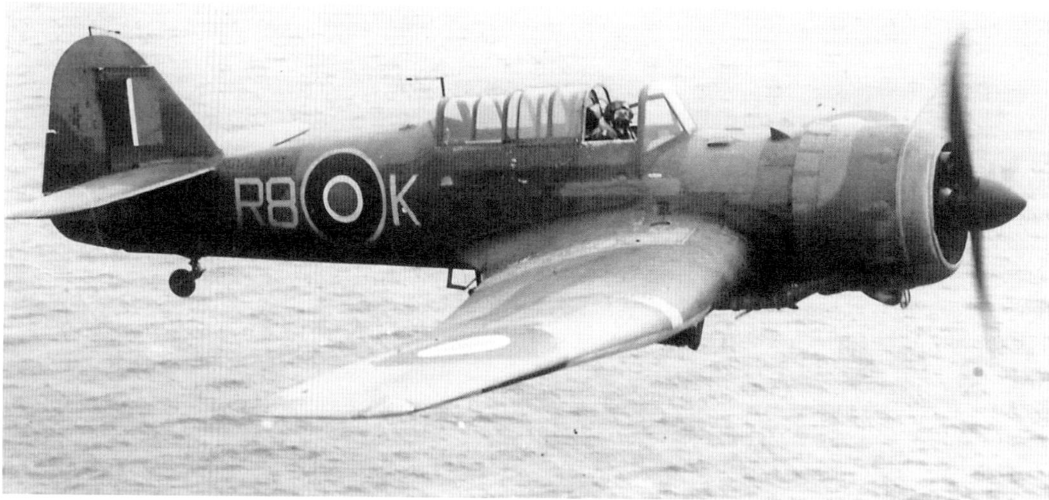

Miles Martinet TT.1 RG988, R8K of 776 Sqn, a similar aircraft to those used at St Merryn and in RAF units in Cornwall

Although there were constant alterations and improvements, St. Merryn was very much as it had been when completed in 1939 – '40. It represented Admiralty thinking which was put into practice in all the airfields specifically designed and built for the Fleet Air Arm throughout the war: four runways 30 yds. wide, one 1,200 yards long, the others approximately 1,000 yards, but all lengths were varied from airfield to airfield. At St. Merryn there were two large and one small dispersals, all with numerous hangars of various shapes and sizes. The technical site was in the south-western corner of the airfield, below the level of the plateau and also had a variety of hangars. As was the case with all Fleet Air Arm air stations, the main function was training, in this case armament training. Fighter pilots, observers and telegraphist air gunners – known as TAGs – did air-to-air gunnery on towed targets, and a range for light bombs was laid out on the cliff-top at Treligga, twelve miles to the north-east along the coast. These activities were incorporated into two second-line squadrons, no. 774[3] Sqn. carried the gunners and no. 792 provided the target tugs. Also making use of these facilities was no. 748 Sqn., a fighter pilot pool comprising Hurricane, Spitfire, Fulmar and Martlet (later known as Wildcat). Pilots were kept in practice while they awaited a posting to a front-line squadron. Invariably a visiting fighter squadron was in temporarily residence for air firing practice.

On our first morning, after the routine of reporting to the Captain's Secretary and the Principal Medical Officer (PMO) we assembled in the office of the Station AEO, Lt. (A) (A/EP) J.Buckley RNVR. The significance of the letter 'P' in his rank description was to recognise his qualification as a pilot. In this instance an important part of his duties was as the Station maintenance test pilot. He had long experience of flying in the Fleet Air Arm and the RAF. He was promoted to Lieutenant Commander (L/C) very shortly after this. In his welcome he told us that we were to be the 1st Air Engineer Officers' Qualification Course. Although we did not know it this was to be the beginning of a period of years of engineering ground training at St. Merryn for both officers and maintenance ratings, finally wresting control of this important function from the RAF.

Looking back, it is disturbing to see how unprepared St. Merryn was to manage a comprehensive training scheme. An overall plan must have been agreed in the Admiralty, because headquarters staff had been appointed to RNAS Lee-on-Solent with Commander (E) Pratt RN as CO. Management of courses for AEOs was transferred from London to Lee-on-Solent. Implementation at Air Station level, at least for our course, got off to a lame start. We heard of the appointment of a L/C (E) (AE) RN as Course CO (an ex-Henlow student) but he never turned up. His successor was of more modest rank, Lt. (E) R.J.N. Duffay RN. A senior instructor was also appointed, Lt. (A) Winsor RNVR. He had been a schoolmaster at RAF Halton, the RAF apprentices' school, and we all thought that his attitude towards us reflected his experience with teenage boys. There was no subsidiary establishment of an administration office or of instructors. The course followed the lines of the long Henlow course, i.e. practical in the morning and theory or lectures in the afternoon or early evening, sometimes in the wardroom. So the Air Station departmental maintenance ratings and their supervisors had to cope with the students for a "course" for which there was no administration and no written schedule or timetable, meanwhile doing their own jobs. The fact that the immediate creation of a routine which mirrored what we had experienced at Henlow was indicative that not much thought had yet been given to the objects and aims of the course. It was inevitable that much of the practical work repeated what we had already done, either at Henlow or at the manufacturers' engine courses. My notes reveal that on that first morning, 14th January, by which date neither Lt. Duffay nor Lt. Winsor had arrived, Lt. Buckley did give us details of the course which was expected to last for four months, but in fact lasted for three. It was obvious that students would be an unexpected "overhead" in each department, but there was much goodwill on the part of the senior ratings and warrant officers who were in charge. None of the supervision could be called instruction. Whatever we were given to do, we had no tools of our own to do it with. Whilst we had been issued with a pair of overalls no hand tools were available. We must have been able to borrow some tools or we would not have done anything.

3 The official designation of a Fleet Air Arm squadron was thus, '800 Naval Air Squadron', abbreviated to '800 NAS'. In the 1940s and '50s daily use of this description was assumed but was abbreviated to '800 Squadron' or, in writing '800 Sqn.' In conversation it was common to omit the words 'Naval' or 'Sqn.' altogether because everyone knew what was being discussed, thus in the case of the arrival of the squadrons at Nowra later in the war, they would be recorded as, for example, '820, 849, 854 and 857 flew their Avengers ashore today'. All our people would know what the series numbers of FAA squadrons would be. RAF squadrons would be recorded or discussed as such; "Typhoons of 56 Squadron, RAF".

It had always been a feature of the system of providing tools for Naval technical ratings that at the start of their training they were issued with a toolbox and a set of tools which they kept as their own throughout their service and guarded carefully. Toolboxes were checked for content every year and any losses had to be paid for. It was the goodwill which made things work for us.

We had no opportunity to discuss the outcome of each assignment and I have no evidence that we were assessed at the end of each one. Two subjects which were exclusive to the students were PT (before breakfast each morning) and parade ground drill on a Saturday morning, for both of which we were in the hands of qualified senior Station instructors. The drill routine, lasting about an hour, started on our second Saturday but the PT did not start until 3rd February; it is depressing to say that on at least one occasion the instructors did not turn up. We were divided into groups of three for our circulation around the Station; I found myself with Jones and Robertson, already friends from earlier courses. All this workshop activity was in the weekday mornings, continuing on Saturday mornings after the drill period.

Naval traditions were not forgotten. We all attended Divisions occasionally; this is the name given to the morning parade which traditionally starts the day in all military services. Sunday divisions were a preliminary part of Church Parade; students were allocated to take command of technical divisions, thus practising their parade ground drill. I enjoyed this. One week of our circulation was spent studying aspects of the management of the lower deck and how punishments and requests from the lower deck were dealt with by the captain and commander.

On most days the afternoons were spent in the classroom watching specialist films or listening to lectures on a wide variety of Service and technical subjects. If there was to be an important visitor to the Station he was given the opportunity to speak to all the Station officers in the evening in the wardroom. One of the major gaps in the lecture series was the history of the FAA and what it actually did now, and how activities at St. Merryn fitted into the scheme of things and contributed to the work of the squadrons in their carriers. We just had to keep our eyes and ears open. By and large the lectures and the films were good. Compared with Henlow, the content and detail of all these subjects were relevant to Naval procedures and equipment. This was a particular bonus and gave us the impression that we really were now in the Navy. Even learning about all the important and relevant publications, forms and procedures was fascinating. Amongst the evening lectures was a particularly interesting and important one by the Rear Admiral Naval Air Stations (RANAS). He spent several days at St. Merryn visiting as many departments as he could and talking to everybody. His evening talk in the wardroom to all officers not on duty was highly confidential and particularly explained future policy for the Fleet Air Arm. One of our regular lecturers was Mr. Storey, the Rolls-Royce field engineer who was based at St. Eval; his talks and demonstrations were very comprehensive and very professional, even if they repeated what we had learnt on our course at Rolls-Royce. The Station was also visited by the new Fifth Sea Lord, Rear Admiral Denis Boyd RN, who was responsible at the highest level for the Fleet Air Arm. He spent four or five days at St. Merryn. On 4th March we welcomed Cdr. Pratt from Lee-on-Solent. He spoke to a gathering of all the Air Engineer trainees but gave little away about the future of officer training, but perhaps this was a learning visit. It was interesting to meet the man who now had the responsibility for creating a structure to train all strata of engineers in the Fleet Air Arm.

Because of all the drawbacks I have mentioned - and others such as the lack of a crane when we wanted one and it took two days for it to become available, a large amount of our time was spent doing nothing. I am sure this would eventually be overcome. One of the departments we visited was Storage (of complete aircraft). Aircraft were continually going into and out of Storage, usually under their own power, so somebody had to taxy them. We were initiated into the starting procedure for a Hurricane and were encouraged to taxy the aircraft around the Storage dispersal – under supervision, of course. I recorded in my diary that the technical activities of two of the squadrons we visited were supervised by Flight Sergeants of the RAF. Even by 1943 we did not have enough experienced senior Naval ratings to take these positions.

I have mentioned that we were introduced to forms and written orders by which the Navy is administered. It was obvious that the traditional system was often incompatible with the operation of aeroplanes. At that time much technical information reached unit level via Kings Regulations and Admiralty Instructions, commonly known as KR&AI. These papers, and their senior Confidential KR&AI were routed through the Captain's Secretary in each unit. More immediate instructions, orders and information were covered by Admiralty Fleet Orders (AFO) and Confidential Admiralty Fleet Orders (CAFO) which were dealt with similarly by the Captain's Secretary. The Instructions covered every aspect of life and work in the Royal Navy. As they were not classified in sections they had to be sorted for distribution to different departments. It always struck me as wrong that such instructions should have to go to the captain's office first. So far as I can remember no communications from outside the Station were ever sent directly to the technical departments, however urgent. Even aircraft instruction manuals, workshop manuals and parts lists, all of which were in sets of Air Publications, and their amendments, arrived in the AEO's office from Main Stores.

Outside our working hours – or even during the day when work had dried up – there were opportunities to watch the flying, and sometimes to participate. Accidents, or "prangs" as they were always called, were inevitable with inexperienced pilots. The trainee engineers were called out twice to put two aircraft, a Martlet and a Martinet, back the right way up after they had turned over on landing. Other accidents regularly involved landing on one wheel after an undercarriage failure. We were very impressed by the reactions of the duty crash tender crews. They always watched runway activities very closely and it was uncanny how the men could see an accident about to happen so that before the aircraft had come to a standstill, whichever way up it was or on fire or not, the crash tender was on its way at full speed across the grass, before the control tower had fired a Very signal. The proximity to the three other coastal aerodromes resulted in the regular unexpected arrival of strange aircraft. The Martinet mentioned was one of those. We also saw a Photo Reconnaissance Spitfire, from Benson, Oxfordshire, en route to St. Eval; a USAAF Bell Airacobra single-seat fighter, probably completely lost; a Harrow twin-engined transport and finally, and the most bizarre, a Handley-Page Halifax four-engined night bomber. This arrival was supposed to have been at St. Eval but, with a low cloud base the pilot had approached St. Eval, set his directional gyro, did a circuit, popped out of the cloud and landed on the first runway he saw – which was St. Merryn, not St. Eval but on a similar heading. There was a screeching of brakes and the aircraft drew up intact. The pilot parked beside the control tower, dwarfing it with the size of his aeroplane. We were interested to see what happened next. A lorry appeared and the Halifax was stripped of all removable items. On a suitably windy day, and with minimum fuel and crew on board, the take-off brought out many St. Merryn residents to see the fun. Holding it on the brakes, the pilot opened up the four Merlins to full power and let her go down one of our shorter runways. Airborne before the end of the runway, the Halifax dropped down into a south-westerly facing coomb at the edge of the airfield to gain speed and clawed its way into the air without harm.

We were encouraged to look for flight experience. I went up in the back of a Fulmar for an air-to-air firing exercise, and, as part of our instruction, L/C Buckley took us up in small groups in the back of an Albacore to demonstrate the requirements of a maintenance test flight.

I recorded the ranks of the department supervisors as we circulated round the various workshops and squadrons. Two squadrons had RAF Flight Sergeants in charge. ERS had a Warrant Air Officer in overall command with additional supervision on the shop floor. Engine overhaul was in the hands of a RAF Sergeant whom we had met on the course at Bristols the previous year. There was no sign of any senior naval technical ratings of any description. It was to be a surprise that we had no RAF men at all in the engineering department at Fearn only five months later.

As I have never been one to look for a friendly pub I spent most of my free evenings in the wardroom. The small guest room was equipped with a good electric gramophone and a selection of classical music records which we could play as we wished. I joined a small group of Station sub-lieutenants for some very enjoyable evenings. The short winter evenings curtailed outside sport, but I took part in weekend games of rugger and hockey and enjoyed some energetic games of squash. I was a regular member of the Station rugger team and, just before I left, I was invited to play for a representative team from Cornish Naval establishments. Unfortunately, bad weather intervened and the fixture was cancelled. I had left the Station before it could be rearranged. A group of us took long walks on Sunday afternoons, either along the cliff tops or the beach, and it was not unusual to be walking for three or four hours. We enjoyed the well-known places and beaches from Trevose Head to Bedruthan Steps. Reading my diary, I am reminded that we enjoyed good weather for all those three months and it was only rarely that we were inconvenienced by rain.

It was during February that I was first informed of a pay rise to come. I think I kept all the official documents concerning my service, but I have none which tell me the date or reason for such a rise. I received a memo from "RN Barracks, Lee-on-Solent" dated 11th May 1943 which recorded that on 1st December 1942 I was confirmed in rank as a "Probationary Temporary S/Lt (A) RNVR". On my original enrolment I was also "Acting". During April the pay rise was implemented. It was for the sum of 4/- per day, bringing the daily rate to 13/-. This might seem inconsequential but, for the times, it was an important sum of money and signalled my acceptance as an established Reserve officer. As with all wartime officers I was still "Temporary". Following our St. Merryn course and qualification we enjoyed a further pay increase during the summer. What had as much significance as anything was that the memo from Lee-on-Solent had come from there and not from the Admiralty. Commander (E) Pratt RN was now his own master.

On the 14th April we took our Course exam, set by Lt. Duffay. I made no notes of the content, but we all passed with over 80%. As from 19th April 1943 we were qualified AEOs. I was told that I was scheduled for another visit to Rolls-Royce at Derby but this was altered at once and, to my astonishment, I was ordered to attend the Air Electrical Course at the SofAE, Henlow, starting on 22nd April. I was dumbfounded. My friend Peter Cox joined me, equally shocked. The Qualifying Course officially ended on April 16th so we were able to squeeze in two days' leave before returning to Henlow. I went home for the first time for nearly six months.

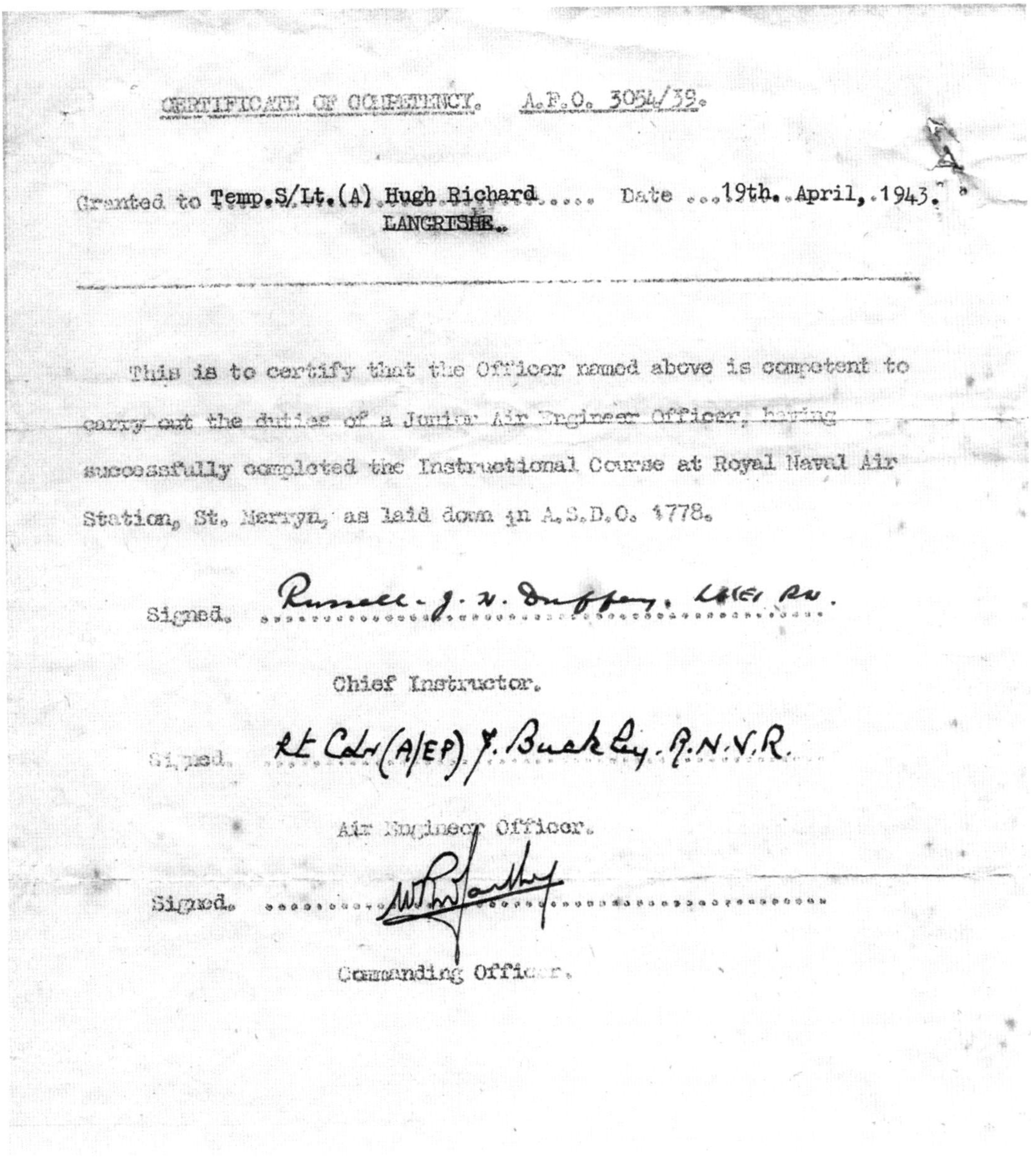

CERTIFICATE OF COMPETENCY. A.F.O. 3054/39.

Granted to Temp.S/Lt.(A) Hugh Richard LANGRISHE. Date 19th. April, 1943.

This is to certify that the Officer named above is competent to carry out the duties of a Junior Air Engineer Officer, having successfully completed the Instructional Course at Royal Naval Air Station, St. Merryn, as laid down in A.S.D.O. 1778.

Signed. Russell-J. N. Duffay. Lt(E) RN.

Chief Instructor.

Signed. Lt Cdr (A/EP) J. Buckley. R.N.V.R.

Air Engineer Officer.

Signed.

Commanding Officer.

Certificate of Competency. A very important document indicating that the person named has passed all courses and can be appointed as a qualified Air Engineer Officer at any FAA establishment. It is promulgated by Admiralty Fleet Order

Fairey Albacore L7083. It was planned that the Albacore torpedo-bomber would replace the Swordfish. This never happened and most squadrons were involved in war on land or from land, either in the Channel against German coastal forces, or against Axis forces in the Mediterranean. The Albacore was fitted with a Bristol Taurus radial engine

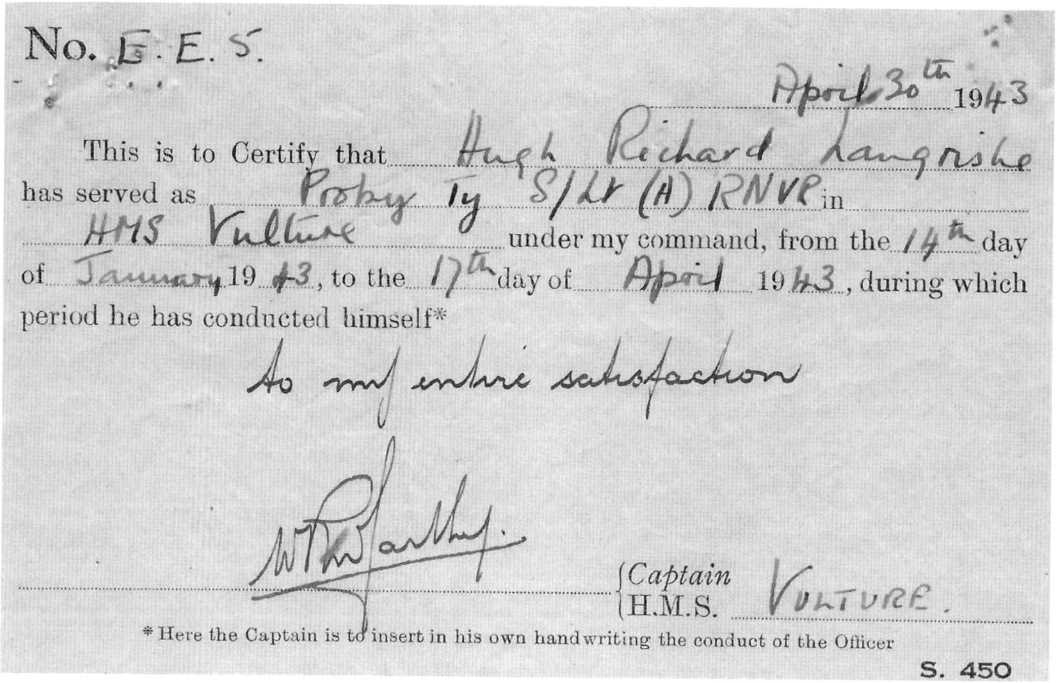

No. E. E. 5.

April 30th 1943

This is to Certify that Hugh Richard Langrishe has served as Proby Ty S/Lt (A) RNVR in HMS Vulture under my command, from the 14th day of January 1943, to the 17th day of April 1943, during which period he has conducted himself*

To my entire satisfaction

W R Carthy [signature]

Captain H.M.S. VULTURE.

* Here the Captain is to insert in his own handwriting the conduct of the Officer

S. 450

Every officer receives one of these certificates on his departure from a Naval establishment, or when command changes during his service there. Unless the person concerned has had regular contact with his captain to the extent that they actually knew each other, and a proper assessment could be made, the flimsy usually contained the phrase "To my entire satisfaction".

A Fulmar II, N4062, of 759 Sqn. Yeovilton 1941. 805 Sqn was the first to operate this aircraft overseas at Dekheila in Alexandria, Egypt.

Air Engineering Training: RAF Henlow School of Aeronautical Enginering, Air Electrical Course

The reasoning behind this course was never made clear. We found that the Naval participants on the course numbered eight, all being AEO Sub-Lieutenants RNVR. Except for Cox, I had never met any of them before. The course was, naturally, specifically programmed for RAF people and was three months long. Subjects appropriate for the Fleet Air Arm would be selected from the normal syllabus, a sort of "pick and choose", so that our course would last for only two months. This led to a rather inefficient use of time. If a lecture was considered of no interest to the Naval group we were just left on our own with nothing to do. This was not just an isolated occurrence; my diary regularly records wasted periods. As with the engineering course, there were serious theoretical lectures. Some of these undoubtedly covered the ground of an electrical engineering course at university and were not out of the way for the RAF officers on the course, most of whom were electrical engineering graduates. I struggled. Towards the end of the course we were visited by a Capt. (E/P) RN and a Cdr. RN who had come to Henlow to see how we were getting on. They interviewed three of us alone and we expressed our concerns about the course as carefully as possible. Nothing happened as a result that affected us, largely, I suppose, because it was so near the end of the course. However, there were repercussions as a result of my general behaviour. Several months later I was summoned to the captain's office at RNAS Fearn and given a dressing-down for my performance at Henlow. Looking back, I probably deserved it. To this day I do not know why young S/Lts (A/AE) were sent on these electrical courses. Six months or so later, when an Air Electrical Officer was appointed to RNAS Fearn, he was not an Air Branch officer, but a S/Lt (L) - Air Electrical – RNVR, with a green stripe on his sleeve. A brief summary of this course is in Appendix C.

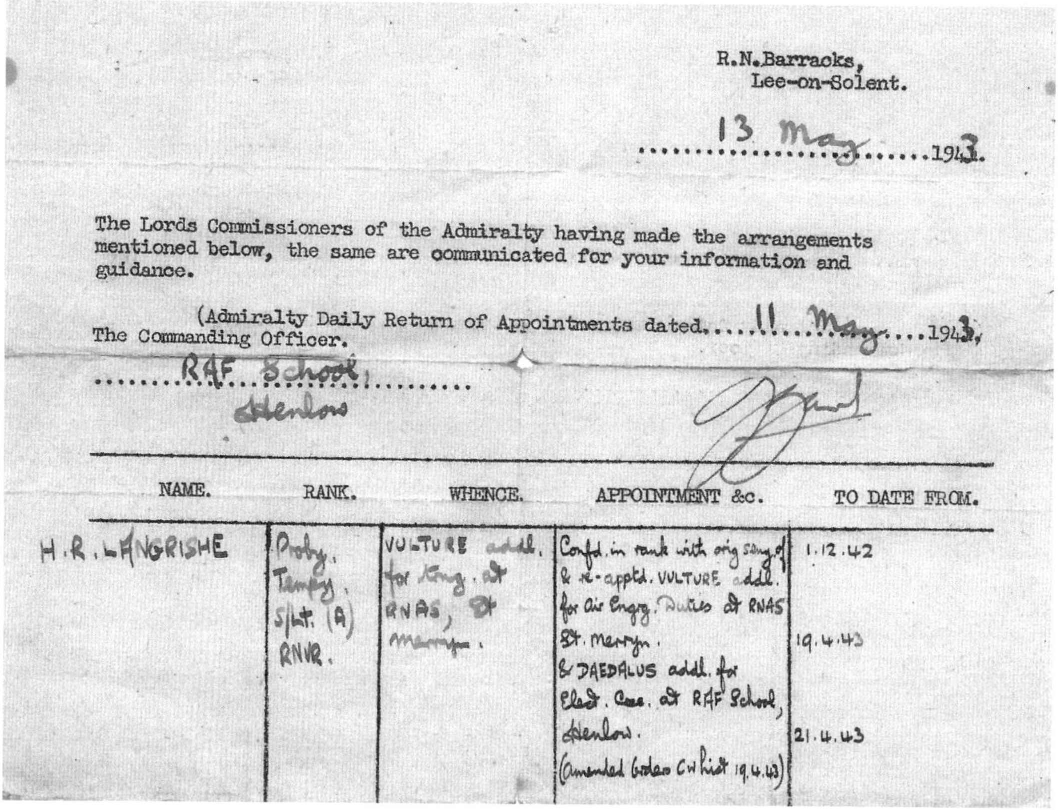

R.N. Barracks,
Lee-on-Solent.

13 May 1943.

The Lords Commissioners of the Admiralty having made the arrangements mentioned below, the same are communicated for your information and guidance.

(Admiralty Daily Return of Appointments dated 11 May 1943,

The Commanding Officer.
RAF School,
Henlow

NAME.	RANK.	WHENCE.	APPOINTMENT &c.	TO DATE FROM.
H.R. LANGRISHE	Proby. Tempy. S/Lt. (A) RNVR.	VULTURE addl. for Engg. at RNAS, St Merryn.	Confd. in rank with orig seny. of & re-apptd. VULTURE addl. for Air Engg. Duties at RNAS St. Merryn.	1.12.42
				19.4.43
			& DAEDALUS addl. for Elect. Cse. at RAF School, Henlow. (Amended Orders CW list 19.4.43)	21.4.43

This document is interesting in that it shows that the function of directing officers to their appointment has been devolved from the Admiralty to RNB Lee-on-Solent. The document also tells me that my probationary status has been lifted so that I became a Temporary Officer, as were all wartime appointments in the RNVR.

Since my previous spell at Henlow something had become a regular feature of No. 14 School of Technical Training. That was the introduction of large numbers of trainee Air Mechanics (Electrical) – AM(L) – for the Fleet Air Arm. On 8th May a Lt(L) RN appeared in the Mess, having been appointed their Divisional Officer. For the church parade on Sunday, 6th June, I was put in charge of one squadron of trainees. I was very impressed to see three units out of four were in dark blue uniforms. On 5th May 18 (Engineering) War Course passed out; they had started in November 1942 and had been the junior course during the last term of 16 War Course. An Electrical course passed out the same day. Successor courses appeared in the Mess on 19th May, each with a Naval contingent. The last day of our course was 16th June, when we had an "Applied Practical" exam. Having had no news of our next appointments, the adjutant telephoned Cdr Bailiff at Queen Anne's Mansions. He said he was not expecting us to finish until 30th June so had nothing set up! We were to take "indefinite leave pending appointment" and we departed for our respective homes. I can find no record of having received the results of the exams we took on the last few days of the course.

At least the social life at Henlow mitigated the disaster of the course. Rugger had finished so there was a lot of hockey and badminton, either after tea or later in the evening as the days got longer. I also attended the Music Club on several occasions. It is worth recording that the domestic side of the Henlow Camp had been well designed and laid out and was very well kept; the architecture of the buildings was to the standard 1930s "expansion" designs used in many RAF stations all over the country. It was unfortunate that so few of the new Naval Air Stations had even been started before the war. Most of them were developed with utility buildings such as the family of Nissen huts. These gave rise to the local naming of many domestic camps as "Nissen City".

Thus it was that after a very lengthy training period, I was at last to be let loose as a qualified Air Engineer Officer. I took up my first post at RNAS Fearn, Easter Ross, on 30th June 1943. I was very excited! No doubt there are quite a few "buts" to be voiced after all that I had experienced over the previous year. I have already indicated that there were mitigating circumstances caused by the short time between the emergence of the Fleet Air Arm from the clutches of the Royal Air Force and the outbreak of war in 1939. But it must have been the case that there were just too few Royal Navy officers with any experience of aeroplanes at all to cope with the enormous task of "navalising" RAF practices. In consequence, the Fleet Air Arm of the period immediately before the war adopted the engineering structure of the Royal Air Force to which it had worked for nearly twenty years. The RAF technical ranks and grades of skill were taken over in toto. The only difference was the colour of their uniforms. To add to the irony, the Royal Navy possessed no Technical Training Schools and large numbers of young men and women continued to be trained in RAF Schools of Technical Training for many years. The effort put into creating an Air Engineer Officer structure by transferring officers from the RN engineering cadre was not altogether satisfactory. Many of these officers brought with them a heap of RN baggage which did not always gel with the character of the young "hostilities only" officers.

I am afraid to say it, but the management of details at Queen Anne's Mansions was poor – wrong dates and duration of courses, for example, are nothing to do with what clothes you have on your back. I suspect that there was no serious investigation of the content of the Henlow courses to see what, if any, modifications should be incorporated to conform to Naval practice – but that would be because there was nobody in authority who understood the problem. After all, if your participation in a training scheme allows you to enter 50% of the students, you must have some leverage. I thought the School of Aeronautical Engineering course was very good and recognised that the engineer officer is not being trained as a skilled tradesman but must have knowledge of the basic requirements of all skills under his command.

One major attribute of Henlow was the close connection with the MU. Visiting the various departments was an especially enlightening and rewarding experience. I think it is important that young trainees are introduced to a complete aeroplane at an early stage in their training so that they can see what they will experience when they qualify – a "live" aeroplane is a serious stimulant.

Fairey Barracuda II. Note the prominent Fairey-Youngman flap/dive brake under the wings, in the neutral position. The wheels retract into the wings and the undercarriage folds into the side of the fuselage and into the wing. The unidentified lieutenant indicates the size of the aircraft

I have given some thought to the use and usefulness of junior air engineer officers. Compare the responsibilities of a squadron engineer to the engineer in charge of an Air Station ARS. The first one may have a dozen operational aircraft and over a hundred technical people to look after, the latter perhaps twenty to twenty-four skilled men and a workshop capacity of eight aircraft in various stages of repair or overhaul. At Fearn I had between eighteen and twenty-four men and women with an experienced AA3 in charge. There were times when I felt that I was seriously under-employed. It surely must be obvious that to be put into a position of authority one should be informed what responsibilities should be shouldered, and how one should do them. Such tools as "job specification" and "scientific management" were never examined at any stage in our training. As regards the first of these things, it might well emerge that a commissioned officer is not always required for a particular post and that a senior rating such as an AA3 or AA2 would suffice. It was a revelation to me when I was introduced to such techniques in my course of production engineering at the De Havilland school after the war.

PART TWO: ROYAL NAVAL AIR STATION, FEARN, EASTER ROSS

Putting training into practice; life in northern Scotland

Where and what was Fearn?

At the time of completion of my training Fearn was unknown in the Fleet Air Arm as an active and vibrant Station. I was soon to understand why. Unlike most of the established Naval Air Stations elsewhere, Fearn was very isolated. There was no town of any size in the vicinity. The Air Station was named after a small scattered village in the county of Easter Ross, northern Scotland, and was only a few hundred yards from the northern, open end of the Moray Firth. The nearest town, Tain, of a size probably referred to as a village in southern England, was some six miles away to the north-west on the southern edge of the Dornoch Firth. Inverness was 45 miles to the south by road and rail, 20 miles by air. I found the surrounding country beautiful. I took my own bicycle with me and I had many wonderful rides and walks along the coast and inland. The narrow coastal belt is very fertile, and crops and livestock thrive, adding to the attractiveness of the area.

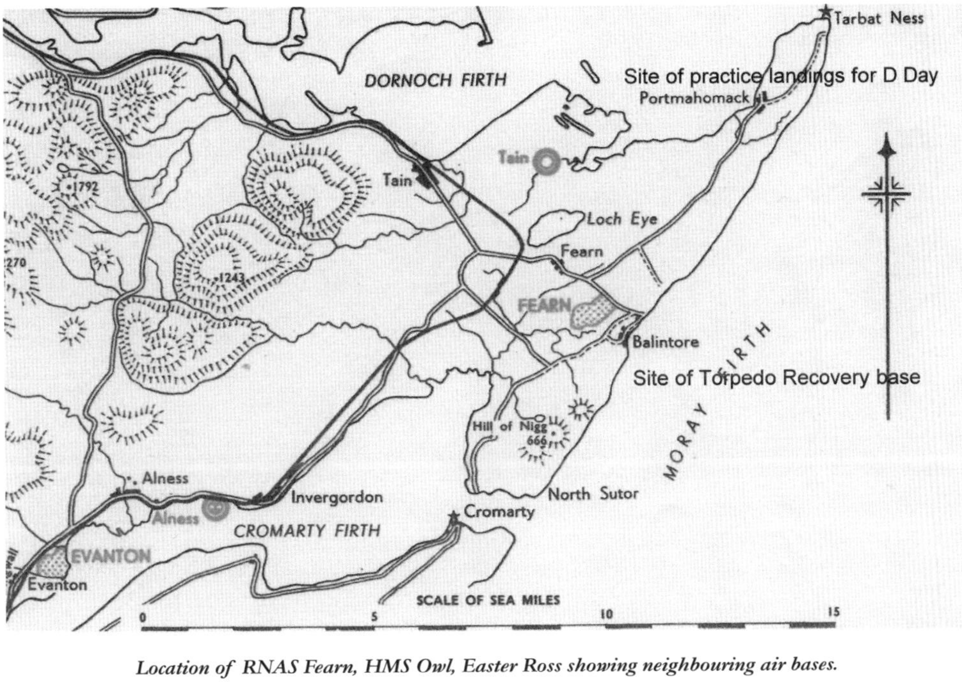

Location of RNAS Fearn, HMS Owl, Easter Ross showing neighbouring air bases.

I had never been beyond Perth in my boyhood in Edinburgh so the journey north on 29th June 1943 was an adventure, although tedious. I think I changed trains at Perth, from LNER to LMSR (the old Highland Railway) and the whole journey must have taken four to five hours. I noted that the train's arrival in Inverness was delayed by 1 hr. 18 mins. due to an incident on the line. The station staff there were not in the least put out and my connection on to the Wick branch departed 1 hr. 10 mins. late. What I saw from the train of aviation activity gave me an inkling of the large number of bases in the far north of Scotland. Inverness itself had a small grass airport between the town and the sea, which, as I learned subsequently, was used by the Dominies of No. 782 (Communications) Squadron from Donibristle on their scheduled service north to Hatston in the Orkneys. The railway passed close to Evanton aerodrome, which was on a spit of land on the edge of Cromarty Firth. It had been used for disembarked squadrons from the Home Fleet in the 1920s and 1930s when the fleet was deployed to the small Naval base at nearby Invergordon during the annual Naval exercises in the North Sea. Its current use was as an RAF air gunnery school, but it also housed a small Naval Aircraft Repair Yard which was responsible to the R.N. Aircraft Repair Yard at Donibristle, Fife. Immediately north of this Station was the flying-boat base of Alness which housed a large operational training unit "(OTU)" for Catalinas and Sunderlands. It did surprise me to see how close these two bases were to Ben Wyvis, a mountain of over 3,000 ft. under ten miles to the west whose foothills loomed over Cromarty Firth. Fearn village and Balintore, a coastal fishing village, had their own railway station known as Fearn, between Invergordon and Tain. Less than five miles north of Fearn airfield was RAF Tain, on the edge of Dornoch Firth. An airfield had been developed beside what had been a bombing range in the immediate pre-war years. It was mainly for the purpose of providing an emergency landing ground for aircraft which might have been in difficulties over the range. When the airfield was developed, the bombing target was moved out into the Firth and a torpedo range was also laid out. As the war developed in the north Atlantic and around the Norwegian coast, Tain's three runways were extended to be able to handle Coastal Command reconnaissance and attack aircraft. Two runways were extended to 2,000 yards to cope with long-range Liberators. After the war, when the airfield was abandoned, the old target area for air-to-ground bombing and rocketry was restored and has been in regular use for many years since.

Consolidated PBY-5 Catalina 1 AH550, DA-L of 210 Sqn, Coastal Command, c 1942. By 1943 many Catalinas had been painted in N. Atlantic white. This model was used at Alness

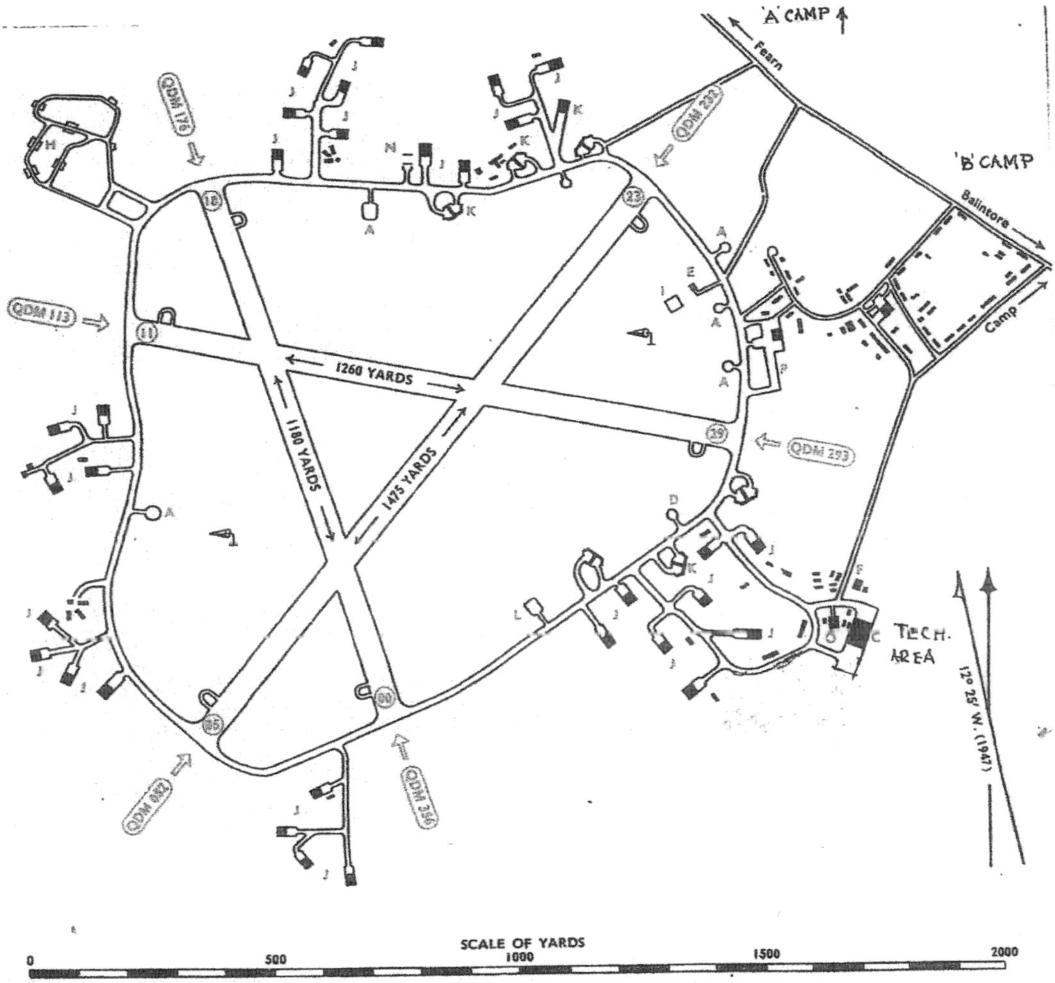

Plan of airfield, RNAS Fearn 1944.

Key:

A	Aircraft standings	J	Hangars
C	ARS Hangar	K	Aircraft pens
D	Compass base	L	Radar test base
E	Flying control	N	Gunnery test butts
F	ERS	O	GES
H	Explosives	P	Torpedo workshop
I	Ground signals		

RNAS Fearn had a very short life in Naval hands. The airfield was originally to have been a satellite of Tain for use by Bomber Command. It was laid out in standard RAF three-runway form, the longest runway, NE/SW, being 1,480 yards. Runway width was 50 yards, 20 yards wider than was normal for an Admiralty-designed airfield. It had a full three-runway Drem lighting system for night flying, including comprehensive perimeter track flush lighting. The whole area was almost completely flat and underlaid by sand and gravel beds which were exploited after the war. This extended to the administrative site – what we called "A" Camp – which was admirable for games' fields. Being low-lying it tended to be boggy, particularly at the southern end where the ground gradually changed to reed-beds and mudflats in Nigg Bay. It was here that the bombing target for Fearn was built, used often by Fearn's Barracudas. Practice with dummy torpedoes was carried out in the Moray Firth, the recovery launches being based in Balintore harbour. The associated torpedo workshop was built outside the north-eastern edge of the airfield with its own access to the perimeter track. It was a very good airfield. Proximity to the sea must have had a beneficial effect in the winter. Even on days of extreme frost, the runways were open when neighbouring airfields were closed. I cannot recall the use of either snow-clearing equipment or anti-freeze chemicals. What may have deterred the RAF was the low (650 feet) Nigg Hill to the southeast. The Fearn anti-clockwise circuit as used by the Barracudas never seemed to be hazardous. Although the Station had been transferred to the Royal Navy in July 1942 and commissioned as HMS Owl, use had been modest. 747Sqn was the first lodger unit. It was formed there in March 1943 as a torpedo, bomber and reconnaissance (TBR) pool but evolved into an OTU with Swordfish, Barracudas and a couple of Ansons for radar training. It never seemed to possess more than four aircraft. In June it moved to the recently opened Naval Air Station at Inskip in Lancashire. It was not until the summer of 1943 that Fearn came into proper use, but after the final squadrons left in May 1946 the place was closed and it was handed back to the RAF in July of that year.

Avro Anson 1 N5331. In 1936 this was the first monoplane with retractable undercarriage to enter the RAF. By about 1940 it had been relegated to training duties but production of the Mark 1 continued until 1945 (a simple modification to the cabin floor introducing the Mark X). Over 700 were built by A.V.Roe., with more in Canada and in Australia. No fewer than 47 Naval second-line squadrons used the Anson 1 for such purposes as training in navigation, radio, radar and air gunnery, including 747Sqn.

At the beginning there were no hangars and no workshops, although by the summer of 1943 a comprehensive building programme was under way. Work on the airfield was not completed until well into 1944, but the additional domestic site known as "B" Camp was ready for occupation by the end of 1943. "A" Camp, laid out some 600 yards to the north of the Fearn to Balintore road, had all the official and ancillary buildings you would expect to find on any Naval establishment: the ship's offices, guardroom, motor transport (MT) yard ,wardroom, mess halls, gym/cinema, squash court and a full range of accommodation huts for every man and woman on strength. There was a very good, flat, sports field on which cricket, hockey and rugby were played. The camp was very compact and convenient if your job was not on the airfield, but MT was hard to come by so to get to work on the airfield you either walked or used your bike. Security fencing for the airfield was by then almost non-existent so a cross-country footpath had been created across a piece of rough and slightly hilly ground to reach the airfield opposite the control tower; understandably this was called the "Burma Road". In view of the small number of men and women who had to get to the airfield, there was no orderly movement in either direction during the day. It looked, and was, shambolic! It took the best part of twenty minutes to cover the distance which was almost a mile. In the domestic camp the clusters of huts were widely dispersed into what were termed "sites".

The officers' sites had several huts, each of which, to start with, had single cabins even for sub-lieutenants. There was a detached ablutions block on each site but the bath house was five or six hundred yards away near the wardroom.

The new "B" Camp was close to Balintore and was enclosed by its own fence. It was on the north side of the public road and opposite the main gate to the airfield, and had a small guardhouse at the gate. It was truly a "Nissen City". The only brick building housed the electricity sub-station for the camp; there was no sports field or parade ground. There was accommodation for junior officers and chiefs, petty officers and ratings and a substantial Wrennery. All sub-lieutenants on the station staff lived in dormitory huts eight or so together but visiting aircrew stayed in "A Camp". Pupil officers attached to the two training squadrons, Nos. 714 and 717 in 1944, lived in "B" Camp while their instructors stayed in "A" Camp. Although the Station's senior officers remained in "A" Camp a new suite of small huts was provided exclusively for the Captain and his wife in "B" Camp. I think it was just before Christmas 1943 that we flitted from one camp to the other. We certainly had our wardroom Christmas dinner there. It seemed a bit of a comedown to revert to communal living after the single and double cabins we had occupied in "A Camp", but by shuffling and arranging furniture we managed to create the appearance of individual spaces. With one exception, our hut was all technical people. It was situated almost at the north-easterly fence, at the highest part of the camp and was placed in a north-south position. Access was through an un-roofed porch at the southern end. From the emergency door at the north end we had an incredible view for miles and miles over Dornoch Firth to the Sutherland coast in the distance. The extent of sky was huge, and later in the winter, night after night, we could watch the Aurora Borealis playing in the northern sky. It was breathtaking. Keeping warm was always a problem everywhere in the camp during the winter. Our hut had one small coal-burning stove in the middle; every night it was stoked up until the whole thing and its flue were red hot so we could have a reasonable degree of heat as we turned in.

At the time of my arrival, development of station facilities was proceeding apace. The programme for hangar construction started in the dispersal in the north-east corner – what was to become the technical site and the main air stores area – and worked round the airfield in a clockwise direction. The dispersals were evenly spaced around the airfield, each having four Admiralty 'S' hangars and associated Nissen huts. At completion there were thirty of these hangars for squadron and storage use and a further three in the technical site. At first only one T2 hangar was built for ARS but a second was built in the summer of 1944. They were almost against the boundary fence on the east side. All sites were connected to the perimeter track by taxi-tracks. For what reason we never found out, the road surfaces in the technical area and the hangar floors were finished in, supposedly, rolled tarred chippings. In fact, there was a totally inadequate amount of tar in the mix, with the result that we were left with a surface of loose chippings, disastrous in an area which was likely to experience dropped oil and other fluids. It was also highly dangerous when running engines on the hangar apron; granite chips were picked up by the propeller and shot to the rear, peppering the underside of the aircraft's fuselage – and anybody who might be standing there.

Short Sunderland III DP200, TA-T of No. 4 (Coastal) OTU, RAF Alness, 1943. The Sunderlands made a majestic sight as they flew close to Fearn every day.

On that sunny afternoon when I was deposited at the wardroom door in "A" Camp I soon understood the workings of the camp. When I suggested to the driver that he could leave me there he said that he would wait and take me and my luggage to the domestic site. This sounded ominous! But it was wise advice and I soon fell into the routine of either walking or cycling back and forth to the wardroom. It was evident that the base as a whole was at an early stage in its development. No. 816 Sqn. was in temporary residence (two weeks) with six Swordfish, otherwise the Station Flight Proctor and Magister were the only other aircraft in use. 816 Sqn. departed for Machrihanish on 8th July. My introduction to the Station the morning following my arrival was equally bizarre. First I was taken in hand by a Lt (A) New Zealand Volunteer Reserve pilot, position unknown, and using a vehicle belonging to the squadron we set off on a tour. He introduced me to the 1st Lieutenant, Lt Cdr Smyth RN, a re-mobilised retired officer with no Fleet Air Arm background, then, on the airfield, to the AEO, Lt (A) J. Crowther RNVR. He had served on *HMS Hermes* early in the war. He was up to that date the only AEO at Fearn. He then took care of me and introduced me to the Captain, Capt. de Courcy Ireland RN and the Commander, Cdr. J. de B. Wailes RN, another officer from the Reserve. I got the impression that the Commander and No. 1 rather approved of the young and lively Fleet Air Arm officers and treated them more like family than junior officers. Likewise, the Army Liaison Officer or

ALO. He was a retired Major of a Scottish regiment. I also got to know Commander (F) Poe RN, a man of Kilkenny background who knew my family there. We had several energetic games of squash.

There was no workshop hangar and no technical department personnel. The AEO had been allocated a small building outside the perimeter track near the control tower which had three rooms at one end and a small garage used by Crowther for his issue Austin 10 pick-up at the other. Also in there was a 350 cc motorbike for our department's use. The Austin pick-up was fitted out with three passenger seats and had a canvas roof. There were a number of them on the station for general light transport duties. Crowther must have racked his brains to think up something for me to do and I wondered if, indeed, he himself had anything engineering to do. It was suggested that I might be designated MT Officer, which was rather out of my remit as an AEO. However, I was given the job of creating an inventory of MT spares which were kept in our garage. Thus I was now an integrated member of the Fearn establishment..

Dreaming of an Air Engineering department

The room I was allocated next to Crowther's was bare of furniture. I was sent to introduce myself to the Stores Officer, the charming Lt Cdr (S) Blunt RNVR, a solicitor in civilian life, who produced two desks, filing cabinet, shelving, cupboard and chairs. Later I was to receive a large selection of stores catalogues and lists. Those relating to aircraft were Air Publications. I came face to face with the remarkable series of illustrated parts lists for each aircraft and engine type and for Aircraft General Spares (AGS). There were also the descriptive workshop and repair manuals for each type of aircraft we could expect to have – in fact, we never aspired to any types other than the Barracuda; we may have held parts lists for the Station Flight aircraft which eventually became just one Stinson Reliant. Several days after my arrival our staff was increased by a Wren who had been trained as a checker, for what purpose in our situation I could not imagine. Much of the next few weeks was spent separating aircraft technical equipment from general stores and moving it out of a Stores' hangar into another temporarily allocated to us. Both of these Admiralty 'S' hangars remained in the hands of our two departments. Some items required a crane which, surprisingly, we had. It became a "friend" during the next few months; although it was the responsibility of the MT department, most of its use was on the airfield or by Stores. It was a Coles crane and I think it had been allocated to Fearn primarily to handle torpedoes. It was unfortunate that its lifting capacity was nowhere near what was needed to lift even one wing of a Barracuda, let alone a complete aircraft. We had runway incidents which we just could not clear up because we did not have the lifting capacity. How we managed, I cannot remember, but it was serious enough to bring out the brass hats to "supervise". The crane's lift and road motion were electrically powered by battery which was kept charged by a four-cylinder car engine and generator. My notes show that I was embarrassingly under-employed in my speciality and spent some time helping Lt Cdr Blunt's people to move stores from one hangar to another. I don't seem to have communicated with Crowther on a regular basis about our work. I cannot recall a regular talk to assess our progress. If he had any correspondence, he must have taken it to somebody else's office, because for several more weeks we had no internal secretarial capacity. We did acquire a Wren writer/shorthand-typist after another month.

Fairey Barracuda II, 736 Sqn. RNAS St. Merryn, School of Air Combat. Type of aircraft in predominant use at Fearn during this time

The day 816 Sqn. departed (8th July) we at last saw what was to be our main responsibility for all my stay at Fearn, the Barracuda II. That evening two aircraft of 847 Sqn. arrived from Lee-on-Solent. The squadron had formed for the first time on 1st June 1943 at Lee with twelve aircraft. Some of the remaining aircraft appeared late in the afternoon of the 13th. Ten had left Lee but two had force-landed en-route, their crew unharmed; the eight had stopped for the night at RAF Church Fenton, then for lunch the following day at RNAS Crail. Another aircraft went unserviceable at Crail so finally seven reached Fearn. Squadron personnel and equipment arrived at Fearn on 12th July after a long train journey from Fareham. The squadron occupied the dispersal vacated by 816 Sqn. where all their aircraft could be kept in hangars recently completed. They were to be with us a month before ten aircraft departed for Machrihanish on 14th August. We were to become very familiar with the raucous noise of the Merlin 32 as 847 embarked on a busy flying programme.

My notebook of the time records many days of under-employment over the months of July, August and September. Looking at the situation from seventy-five years later, I am surprised that it was like that. We did not lack VIP visits so there must have been exchanges of view with the Station staff, although I cannot recall if anybody asked the young "subbies" what they thought. The first visit I recorded was on the first Sunday of July, less than a week after my arrival. It was by A/Capt. Tyndall-Biscoe RN, the Asst. Dir. of AMR, which sounded something like "Aircraft Maintenance and Repair". On the following day, 5th July, a four-engined DH86A of 782 Sqn. arrived from Donibristle with the RANAS (N) and his staff, one of whom, his Air Gunnery Officer, was Lt Lemon RN, whom I had last seen when he was Air Gunnery Officer (AGO) at St. Merryn. The last visit I recorded was of the 5th Sea Lord, on 30th August. We understood that he had come for a face-to-face talk with the captain. The visit was only for the afternoon. The party had come in a DH 95 Flamingo of 782 Sqn., Donibristle. It was a very rare aircraft, the only one of its type used by the FAA for communications around the Scottish Naval Air Stations. There was another visit much later. It was after I had stopped writing a diary, so I have no date for it. It must have been important because the "chippies" had to build a saluting platform and we paraded formally on one of the runways. I was put in charge of the technical division. We must have done some drill practice because the event went off well. Afterwards I was commended for the way I had conducted my part of the parade, even to the extent that somebody asked if I had been on a drill course with the Royal Marines! In fact I had been introduced to parade ground drill by a retired Sergeant Major at the age of eight at my first boarding school and I had always enjoyed the precision of it.

Looking back seventy-five years, I can see that the explosive expansion of the Air Engineering branch at that time was badly managed. In brutal terms, it was obvious that the branch of the Fleet Air Arm responsible for planning just had neither the experience nor the knowledge to do its job. The young men sent out to the Air Stations were not instructed in how to set up and manage a workshop. Indeed there was no breakdown of tasks or responsibilities – what I would later know as "job specifications". Until they arrived, I did not know what would be the numbers of men and women I was allocated for my department, nor what their rates or specialities would be. All we knew about the equipment and work of the various departments was picked up at St. Merryn during our qualifying course, and this knowledge was not formalised in the classroom. In my case it was not to be before October or November, when the first Type T2 hangar for ARS was completed, that I felt that I was doing a proper job. Even then, I was reliant on the Stores department to provide a range of equipment for which only they had a schedule. The gaps I found were very difficult to fill because Stores had to refer to their own superiors at Perth to have my requests ratified and the schedules altered. I had to rely heavily on my "Chiefy" to guide me in the day-to-day management of the workshop and the allocation of the different levels of mechanics and fitters.

De Havilland Flamingo. One aircraft, DT312, was used by No. 782 Sqn. Donibristle, Fife, for communication flights by RANAS(N) around the numerous Naval Air Stations in Scotland. It visited Fearn once during my time there.

This was a time of rapid development and expansion of Air Engineering activities. In July I heard from one of my Henlow course friends, S/Lt Peter Cox, that he was about to take charge of a Special Repair Party (SRP) for the Pegasus and Taurus aero-engines. He was doing courses at Bristol's. He told me that two other fellow-trainees, Tryhorn and McCulloch, had gone to the States to learn, separately, about the Pratt & Whitney and Wright aero-engines used in American aircraft before forming SRPs. Harry James (15 Course at Henlow) was forming a Barracuda SRP and Roger Ginger, also of 15 Course, a Seafire and Spitfire SRP. I was to see more of James and his Barracuda SRP in the middle of August. He turned up at Fearn in a Ford Utilecon van on 15th August, leading three lorries carrying the men and equipment for what was now called a Special Servicing Unit for the Barracuda II. He was due to be attached to 823 Sqn. which transferred from Lee-on-Solent on 19th August. 823 had recently completed a busy time on the south coast from September 1942 to June 1943, attached to Coastal Command, for night anti-submarine patrols, flying black-painted Albacores from Manston and Tangmere. They had returned to the FAA in June '43, re-equipping with the Barracuda II at Lee-on-Solent. Fearn was to be their working-up base until January 1944. I got to know their aircrew very well in the "A" Camp wardroom and went flying with them on several occasions. One of these was dive-bombing practice on the Nigg Bay target with S/Lt Mac Rutherford DSC. Our paths crossed again in 706 Sqn. in Australia in 1945 and finally when he was CO of 1832 RNVR Sqn. at Culham in 1950.

It is clear now that by 1943 there were many Sub-Lieutenant AEOs qualifying and spreading through the FAA. During August and September 1943 Fearn received a S/Lt Air Electrical Officer (not (A) but with a green stripe) and two S/Lt AEOs for Station Air Engineering duties; the latter arrived when Crowther was on leave and he had left no instructions for their employment. I wonder if he knew they were coming. During the days and weeks of that long summer, when I tried hard to find things to do, I made the most of the opportunities to increase my driving experience on as many different vehicles as I could. I had a civilian "Learner" licence and I must have convinced somebody at Fearn that I was competent enough for a Service licence. I tried my hand at the Ford Utilecon, a mini-bus based on the Ford 10 light van (known as a Tilly) the Austin 10 pick-up with the three canvas seats as well as the driver's; a Bedford 3-ton 4 x 2 lorry; a Bedford articulated 3-ton truck and trailer built for carrying torpedoes, and the Coles crane already mentioned. Later we received a Fordson tractor for towing aeroplanes. I had become familiar with motorcycles when I rode on the pillion of my Edinburgh friend Ron Flockhart's bike at university but I had never had the opportunity to drive a machine myself. However, the presence of the bike in the garage next to my office was temptation indeed. On many evenings, after flying had finished for the day, I would take the bike out on to the perimeter track, practising gear-shifting and handling around the airfield dispersals. For some reason we had also been issued with a "combination", a 500cc side-valve Norton with a box sidecar. The driving technique is quite different to a "solo" but for all normal manoeuvres I became proficient. Many evenings and at weekends I went out, either on foot or on my bicycle, to explore the countryside; I would cover as many as thirty miles on the bike and went as far afield as the upper end of Dornoch Firth or, in the other direction, Invergordon and Cromarty Firth. On many occasions I walked up to the top of Nigg Hill to enjoy the stupendous view.

RNAS Fearn was a community in sparsely populated country. It was important to keep everybody happy when off duty. The Sports Officer, S/Lt Rae RNVR, who was also the Station Schoolmaster, was a specific appointment. He was a PT schoolmaster in civilian life and he had been a playing member of the Galashiels RFC, the famous "Gala" borders rugby club. He was a very good organiser, setting up inter-departmental competitions in football, hockey, cricket and rugby, and matches with RAF Tain in football, hockey and rugby. There was also an annual athletic sports day in the summer. The departments participating in these competitions were Ship's Company, Royal Marines, WRNS, Chiefs and POs, Wardroom, visiting and/or resident squadrons and Torpedomen. It was sometimes an effort for some departments to raise a team. I noted that on occasion I was roped in to make up the numbers in the chiefs' or Wrens' teams.

Indoors there were regular Station and departmental dances, usually raising money for a Service charity, also film shows and ENSA live performances (of varying standards). Until "B" Camp was opened towards the end of 1943, all these indoor activities were held in the "A" Camp gym/cinema. "B" Camp had a very large Nissen hut for this purpose and it became the main venue for Station events. Within ten days of my arrival there was a big Wardroom dance, using a number of buildings around the wardroom, all imaginatively decorated. Guests were invited from RAF Tain and other local military establishments, including the mountain warfare school at Invershin. It was a great success. The young sub-lieutenants were also in demand to swell the numbers in evening social events at local houses. One evening Cdr Wailes took a car-load of us to a dinner party at the home of a local landowner at Geanies House, only a couple of miles up the coast road. The party was organised for their eighteen-year-old daughter who was about to start an arts course in Edinburgh. It transpired that she had organised "digs" only a hundred yards away from where my family lived. We were to visit Geanies House again not long afterwards to celebrate the daughter's eighteenth birthday, enjoying an evening of games and Scottish country dancing. From time to time a small group of us would take a Tilly to Tain, either on Saturday afternoon to shop or on a weekday evening for a rare visit to the Royal Hotel in Tain for a drink. Also, I was invited by a friend of my sister's, who was an ambulance driver for the First Aid Nursing Yeomanry, attached to the RA coastal battery at North Sutor, overlooking the entrance to Cromarty Firth, to an ENSA[4] concert in their cinema. We exchanged visits on our bicycles.

Grumman TBF-1. Avenger 1 FN766. Delivered to the UK in January 1943. At first sight a great big clump of metal, but it became a well-loved aeroplane among the TBR boys. Funnily enough it was never used in its Royal Navy service to drop a torpedo although it was used successfully in this role by the United States Navy in the Pacific theatre. I had my first close-up of the Avenger when one visited Fearn.

The airfield was busy, but without my participation in any engineering work. 823 Sqn suffered two undercarriage collapses. The entire aircraft were dismantled and collected by low-loader for transport to Fairey's for repair. We had several emergency landings by strange aircraft, both with engine trouble. One was a Martinet target tug from Evanton; the squadron commander came up with an engine fitter and departed in the aircraft shortly afterwards, muttering "pilot trouble". Another was a PR Spitfire which was suffering from high engine temperature and low oil pressure. The pilot was happy to wait until things cooled down whereupon he started up and departed for Inverness, gaining height quickly so that he could glide down to land and avoid trouble en route. We received visits from a Hellcat, a new Naval fighter from the USA, on test at Boscombe Down, and an Avenger torpedo-bomber, which was interesting to compare with the Barracuda. I thought it rather large and unwieldy at first sight although my later experience gave me respect for the type. Visitors from Lee-on-Solent came in a Beechcraft Traveller, a luxuriously trimmed four-seater biplane with a retractable undercarriage, and a Vought Kingfisher. This was a wheeled-undercarriage version of the floatplane I had encountered at Lee. On several occasions we were treated to a "beat-up" by a Mosquito; I learnt that the pilot had a girlfriend in the WRNS at Fearn and one day he actually landed so that he could see her. The meeting must have been approved! I was delighted to have the opportunity of a close look at the aeroplane and what I saw I approved of. I was called in to help start the machine for departure, a job which was made difficult due to the Ki-Gas engine primer pump being on the underside of the starboard engine out

4 ENSA was the Entertainments National Service Association set up in 1939 to provide entertainment for British armed forces personnel during WW2.

of sight of the pilot. Somehow, I managed it without over-priming and without losing my cap when the engine fired. We were given a short display at low level before the pilot departed for Wick, where his squadron was based.

In this memoir I have not yet mentioned that I had let it be known to Cdr 'F' that I considered myself an expert in Aircraft Recognition (AR) and had already lectured on the subject to the University Air Squadron, at Henlow and at St. Merryn to the qualifying course. I was duly recognised as the Station AR instructor. In the FAA this subject was in its infancy. A special department was only just being equipped at Yeovilton to instruct the pupils at the Fleet Fighter School there. An 'A' branch officer was appointed to oversee its development and plan courses of instruction. This subject had been a hobby of mine since boyhood and I had a considerable amount of descriptive and pictorial information taken from current magazines. During the war *The Aeroplane* weekly magazine had started another periodical called *The Aeroplane Spotter* to which I had subscribed since its first issue. I could base my lectures on extracted drawings, silhouettes and statistics. No official syllabus existed and I had to devise my own, matching the requirements of the classes I took. I appreciated that the TBR squadrons passing through Fearn would eventually end up in India, Ceylon or the Pacific, and I had to find my own data and pictures of Japanese aircraft, for want of literally none being available from Naval sources. In the summer of 1944, I was allocated a Nissen hut specifically for Aircraft Recognition. I was able to obtain a number of the 1:72 models available, and a selection of the silhouettes which were now being published officially. I gave lectures to the visiting squadrons and to our detachment of Royal Marines who were responsible for the airfield anti-aircraft defences. I can say that throughout my service in uniform I never met another person who was as clued-up as I was in this subject. My experience during my work with Main Stores had introduced me to the two principal official publications on aircraft recognition. AP1480A was illustrated by silhouettes of all the current types and marks of Allied aircraft in service and was allowed open circulation. AP1480X had a circulation limited to Cdr (F) and the people in the control tower. It illustrated the current experimental aircraft flying and was closely guarded on ordinary airfields like ours. My position as AR instructor enabled me to visit the control tower from time to time to examine the publication. It was arranged that I should visit Yeovilton to see what I could learn from their new AR department. It was a long journey via London, a journey which I subsequently thought rather pointless as there was nothing at Yeovilton which was of any use to me or, in detail, that I did not already know. In terms of practical aeroplane watching there was always something going on at Tain. Detachments even as large as squadron size came up from the south either for practice on the bombing, rocket projectile or torpedo ranges, or for the occasional raid on the Norwegian coast or coastal convoys. I saw a Hampden torpedo squadron in the area for a few days' practice; a Beaufighter squadron came for rocket projectile practice, then the North Coates Coastal Command Beaufighter Wing brought all their thirty-odd aircraft for an operation in the North Sea. The sight and sound were impressive.

One afternoon Harry James asked me if I would like to accompany him in his Tilly to the Royal Naval Air Repair Yard (the naval equivalent to an RAF MU) at Evanton to borrow or otherwise obtain some piece of equipment for his servicing unit. This was a good opportunity to see what went on in a repair yard, particularly in such an isolated place. We found several S/Lt (A) RNVR AEOs whom we had not met before. I made no notes about the experience, so I don't think we were all that impressed with the place. On the way back, and out of curiosity, we found an excuse to visit the flying-boat base at Alness, which we found fascinating. We saw maintenance taking place on Catalinas and Sunderlands, both on land and on the water. It was interesting to see how the mechanics secured their hand tools to a belt by cords so that they did not lose them in the sea if they let go by mistake. It was an instructive afternoon.

I closed my notebook on 18th September. I gave no reason, but I suspect that it was mainly because there was almost nothing to write about concerning my real job. It is a pity that I did not persevere because I have no exact dates of the completion of "B" Camp and the hangars in the technical area of which I was to be a part. We were certainly installed in both well before Christmas. How I came to be in charge of ARS I do not know. It may only have been because I was the first S/Lt AEO to be appointed to Fearn. A lot happened at once. Crowther left us and was replaced by a Lt Cdr(E) RN. His name has gone, which is a pity because he was to guide me for the next eight months. Shortly after his arrival he acquired his "brass hat". It was unfortunate that he rather liked the bottle and I don't think that could have done his subsequent career much good. His appointment was the first step in developing a proper air engineering department. Shortly afterwards we received Lt(A)(AE/P) Jimmy Barrett RNVR as Senior Engineer and maintenance test pilot. Cdr(E) was a newcomer to aircraft but Barrett, although in his late twenties or early thirties, had seven or eight years' experience as a pilot, part of which was gained in Canada before the war. He came to Fearn from Egypt where he had been a maintenance test pilot at RNAS Dekheila[5]. He was a motorcycle and sports car enthusiast and somehow had managed to bring home an Italian army motorbike which somebody had "rescued" from the desert. It was a Benelli 500cc overhead valve twin-cylinder, a very sporting machine. I seem to remember that it came to England on one of HM's ships. Situated where we were Barrett used it every day, uninsured and untaxed, both in and outside the airfield. I also rode it on the open road on several occasions. There were ways and means of keeping it fuelled. I suppose its obvious military appearance gave it some legitimacy. The technical department officers, apart from Barrett, were all sub-lieutenants, three AEOs and one Electrical officer.

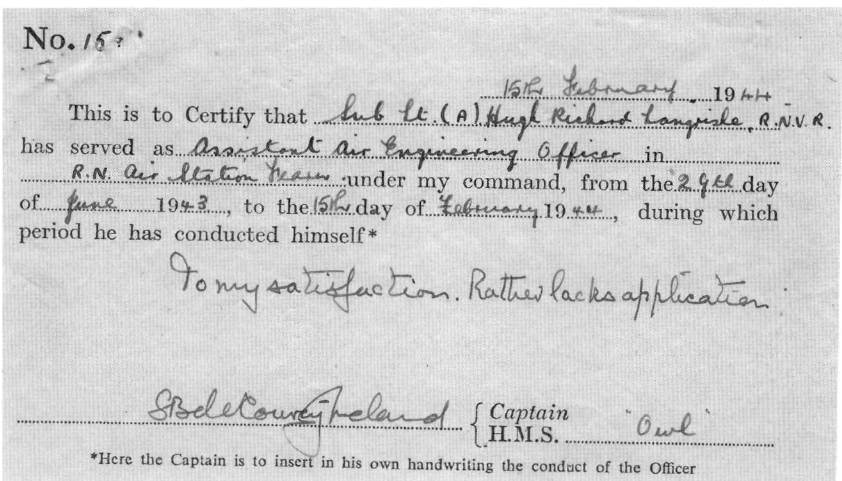

No. 15

15th February 1944

This is to Certify that Sub Lt. (A) Hugh Richard Langrishe R.N.V.R. has served as Assistant Air Engineering Officer in R.N. Air Station Fearn under my command, from the 29th day of June 1943, to the 15th day of February 1944, during which period he has conducted himself*

To my satisfaction. Rather lacks application.

S B de Courcy Ireland { Captain H.M.S. "Owl"

*Here the Captain is to insert in his own handwriting the conduct of the Officer

It was probably inevitable that there would be a negative aspect to this flimsy. All officers in the ship would receive a flimsy on the departure of the captain in a routine manner. Bearing in mind that for the first five months of my posting at Fearn I had no proper air engineering job to do, "rather lacks application" would probably be an appropriate description of my behaviour. The winter of 1943-44 was for me a period of learning on the job.
Capt. Robertson's flimsy eight months later on Page 58 tells a different story.

5 Dekheila had originally been the airport for Alexandria. At outbreak of war it was handed to the RAF who promptly lent it to the Navy and it was commissioned as *HMS Grebe*. It was the main FAA base in Egypt throughout the war. Carriers using Alexandria disembarked their squadrons to Dekheila and it also housed a Fleet Requirement Unit (FRU), a flying unit for cooperation with Fleet Air Defence – possibly where Jimmy Barrett was stationed when he was in Egypt. We also used an airfield at Fayid in the Canal Zone for a repair yard which may also have been Barrett's base.

The technical area comes to life

I can recall the excitement of finally moving into a real hangar in which we could work on real aeroplanes. ARS was a standard T2 hangar, a rectangular building with 18 ft runner doors at both ends and a pitched roof. At the back of the hangar was a modest hut which we used for admin purposes. I had a small room for myself where I kept a progress board on the wall. The major part was used for a store for ready-use materials and parts and for a place for the staff to leave coats and equipment and drink their tea. There must also have been a washroom and lavatory. It is likely that we had one stove and a ration of coal, quite inadequate to provide warmth in the winter. The hangar roof lighting was not really powerful enough to work by and we had too few portable lamps. There were no windows in the sides of the hangar so to obtain daylight the hangar doors had to be opened. It would be interesting to know if the specification of airfield buildings for airfields in northern Scotland took account of latitude and daylight hours. There was no dedicated crane for removing propellers, engines or major airframe parts so there had to be room for the Coles crane to manoeuvre in the hangar. Whatever we were doing we always had the bugbear of dripping engine oil, and there was the eternal problem of how to drain an engine of its oil without getting a fair amount of it on the floor. The Merlin was only slightly better at keeping its oil in than the American radials I encountered later. The state of the tarred and gravelled floor made it impossible to use fluids to break down the oil and gather it up and it was, literally, months before we could get large drip trays from Stores. The state of the floor became an embarrassment. Somebody, either at Lee or in London, decided the numbers of personnel we should have and, because of the system of managing these men and women, for courses or promotion. As if they were all cogs in a huge nationwide machine, there were frequent postings in and out, even after only a few weeks.

Around the turning and running-up pan outside my hangar were two Admiralty "S" hangars, one for ERS and instrument servicing and one for GES which included a "chippies" shop, a fabric shop and probably a machine shop. The radio mechanics, all being Wrens, worked in their own "bubble" in their own Nissen hut beside our hangars. The two chippies were under my control. They were "hostilities only" ratings and

Percival Proctor I P6062. This aircraft was extracted from a batch of Proctors (P5999 to P6113) supplied to Admiralty contract in May 1940 and retained by the manufacturer for their use. It represents the type first issued to the Station Flight, (which was a Mk.II) at Fearn in which I had my first flight over northern Scotland.

behaved just as if they were in their peacetime carpenters' shop. Naval routine was anathema to them. They had virtually nothing to do with aeroplanes, so their work was almost entirely on items for the camps and for recreational activities. For example, when a dance band was formed to perform at Station dances, the chippies made the music stands. They also made a very impressive saluting dais. Good timber seemed to be readily available from Naval Stores, for which Lt Cdr Blunt issued me with a copy of the standard naval stores Parts List. My relationship with the two "chippies" was very much on a peacetime basis and after a while we slipped into Christian names. One day this came to the attention of Cdr (E) and I was given a sharp talking-to: How do you retain Service discipline with Christian names? Mmmm.

In terms of their trades the manning of ARS was consistent for the next eight months although numbers varied between eighteen and twenty. The Chief was AA3[6] Clark who, I think, had been trained in the RAF apprenticeship scheme either at Halton or Cranwell. He had been in uniform for at least ten years and was a good technician and leader. His visible rate was Chief Petty Officer. Next to him were AA4 Claricoates, (PO rate) two PO Air Mechanics (A) and one PO Air Mechanic (E). There was one LAF (A) and the remainder were a mix of Air Fitters and Air Mechanics, including a Wren. It is obvious that manning was directly related to the amount of airframe or complete aeroplane work we were expected to do. If any engine work was to be done in situ, men were sent across from the ERS hangar. Normal routine included Major Inspections (squadron aircraft), carrying out modifications, and doing repairs beyond the scope of squadron maintenance. Although Storage had an AEO and a small team of ratings for routine work, which they could carry out in their own hangars (all aircraft were under cover), major modifications were carried out in ARS. Our men were all trained to do re-sprays and although we had all the appropriate equipment, there was no spray booth or curtain, so we had difficulty isolating aircraft. This was never overcome.

RNAS Fearn ARS personnel, May 1944. Back row: (unknown), AF(A) Vickerage, AF(A) Legg, AM(A)(unknown), AF(A) Smith, AF(A) Cooke, AM(A) Donougher. Middle row: PO AM(E) Aldred, AF(A) Ward, AF(A) Harding, Wren AM(A) Sumner, LAF(A) Heads, AF(A) Farquhar, PO AM(A) Gibbons
Front row: PO AF(A) Ball, AA3 Clark, S/Lt (A) (AE) H.R.Langrishe, AA4 Claricoates, PO AM(A) Alexander

6 AA stands for Air Artificer, 1st to 4th class, AA1 to AA3 rate as CPO (Chief Petty Officer), AA4 as PO (Petty Officer)

Not long after we started work in our new hangar, we were confronted with an urgent manufacturer's modification which necessitated grounding all Barracudas until modified. There had been several reports of the crumpling of the leading-edges of both wings over about two feet at mid-span during vertical or near vertical dives. Drawings and instructions were distributed to all Barracuda units and Stations. There were two parts of the instructions: one in the case of actual crumpling, the second to be a remedial measure in the event of no visible damage. In both cases the objective was to increase the thickness of the skinning of the "D" section of the leading-edge of the wings. Precisely positioned by the drawings, holes had to be cut between the wing ribs in the underside of the wing, in front of the main spar, large enough to get a hand in and to insert the reinforcing metal; this was riveted to the existing skin with blind rivets. Then a closing patch and overlap fillet had to be cut from new metal and riveted in place to both the patch and the wing skin. All new materials had to be properly prepared with the appropriate primer and undercoat and the outside finished in the relevant camouflage colour. It must have been to our relief that we found no instances of crumpling on the wings of 823 Sqn.'s aircraft. Because of the amount of drilling to be done to enable the pieces to be riveted together we needed the speed of compressed-air drills. All we had were electrically powered so where were they to come from? At Cdr (E) level contact was made with the Repair Yard at Evanton who were able to lend us half-a-dozen drills and some high-pressure hose and connectors. We also borrowed a compressed-air nibbler to quicken the process of cutting the sheet metal. The only way of collecting these tools from Evanton was to send me over on the Norton box sidecar motorbike! With some twenty aircraft to do we had to organise the men into teams and work extensive overtime at night. There was great enthusiasm, and the work was done in good time, to the background of the relentless scream of the air tools. I cannot remember how we satisfied the Aeronautical Inspection Directorate because I am sure that this work was serious enough to warrant third-party inspection. When all was done I had another run to Evanton on the Norton to return the tools. There is no doubt this episode was serious for Fairey's. In the aftermath it is easy enough to criticise. In fact, there was no experience of dive-bombing in either the Fleet Air Arm or the RAF and no need for vertical or near-vertical dives of any duration in normal combat. It was in order to clear the elevator from the aerodynamic effects of the Fairey-Youngman external flaps/dive-brakes that the tailplane was raised from a conventional position to high up on the tailfin. Obviously the Fairey Stress department got their figures seriously wrong. Other problems emerged, as explained later. By the end of 1943, I understand that all dives at an angle required for dive-bombing were begun with the dive brakes in the dive position thus restricting the dive speed from the very start; they were then neutralised before starting to level out.

It is rather obvious that AEOs should have a desire to fly. I got to know the Station Flight pilot – each one in turn as they moved on – and my first experience was in a Percival Proctor on a local flight, to make sure the aircraft was working properly. We went out over the Moray Firth and it was a great experience to see the stupendous view all round from a couple of thousand feet. Later I was to fly with 823 Sqn. on a dive-bombing exercise, an experience I did not wish to repeat. I was also taken up for an evasion exercise when the pilot did everything except turn the Barra upside down - climbing, diving and turning to the limits of the aircraft. I also took several flights in the final Station Flight aircraft, the Stinson Reliant, and I once took the 782 Sqn. Dominie service to Inverness to catch a train south. The small grass airfield was only a few hundred yards from Inverness railway station. One of my last Barra flights nearly ended in disaster. We had done a propeller change in ARS and Jimmy Barrett was to test the aircraft after a successful ground run. I was, as usual, in the observer's seat in the "conservatory" amidships, fully equipped with radio headset. Immediately we were airborne Jimmy shouted Mayday. I could see him up above me moving rapidly from side to side. He told me quietly that there was a massive oil leak from the front of the engine, but that as the engine was giving full power and the instruments were still working properly he would gain height, do a circuit and hope to do a normal approach and landing. There was oil everywhere. It quickly spread along the central windows but Jimmy kept everything going and we landed safely, stopping on the runway. When he climbed down, I could see that his head and shoulders were completely covered in oil and it was extraordinary that he had managed to complete the landing. The fault was found to be the failure of the seal which guarded the flow of full-pressure engine oil to the Rotol propeller-adjusting mechanism.

ARS was sometimes used as a depository for damaged aircraft awaiting assessment, usually from crashes or other mishaps. It could be depressing, as at least one tangled mess contained human remains. The aircraft were all retrieved from the marshes of Nigg Bay, having crashed during dive-bombing practice. In most cases the wrecks were taken away at once by low-loader for examination elsewhere, but not always. It was not clear at the time what might be the principal contributor to the inability to pull out of the dive. Late in 1944 some very hazardous test flying led to the discovery that if the dive-brakes were neutralised during pull-out, disturbed air was washed over the tailplane and elevator, breaking down the airflow. The pilot would immediately lose control, and the aircraft would flip on to its back and crash. One such crash ended up in ARS and we were asked to dismantle it in preparation for collection by a "Queen Mary" [7]; I was sitting in my office when a white-faced young man came in holding a flying boot. "What shall I do with this, sir?" he asked. Inside the boot was the leg and foot of one of the dead crew.

There was an irregular turnover of squadrons. 823 Sqn. left in January 1944. 717 Sqn. returned from Inskip, bringing their small collection of Anson, Barracuda, Swordfish and Walrus and stayed for six months. During this time I scrounged a flight in the Anson. The string of front-line squadrons continued with 822 Sqn. on 16th January, for two weeks, 841 Sqn. from 29th February to 28th June, and 828 Sqn. from 14th April to 5th July. At the beginning of July Fearn had a new role, operational training of Barracuda crews. On 1st July 717 Sqn. was formed followed shortly by 714 Sqn. on 1st August. Each squadron had two flights of fourteen to eighteen Barracudas, making a most extraordinary sight when both squadrons were active. Pilots, observers and TAGs came from their individual flying schools, the first two already commissioned, and were "crewed-up" for the first time. The atmosphere was of tremendous enthusiasm.

There were surprisingly few accidents that I can remember, but one stands out. There was still great concentration on torpedo dropping and attack, and almost every day aircraft were out over the Moray Firth, with or without dummy torpedoes. One fine summer day I was outside the north end of my hangar, from where I could see the aircraft approaching to land, flying past with flaps and undercarriage down, no more than a couple of hundred yards away. An unusual high-revving noise became apparent and a Barracuda swept past. A few minutes later there was a call from the control tower asking for a tractor to tow away an aircraft which had stopped on the runway. Gathering tractor, driver and towing arm, I drove the department pick-up over to the runway. I was surprised by what I saw. The prop blades had jagged tips, having lost at least a foot in length, and the underside of the fuselage appeared to have struck something. A word with the pilot explained it all. He had flown too low on his dummy torpedo run (no torpedo) and had hit the sea and bounced back into the air! His luck was still with him when he came to lower the undercarriage and flaps to land and everything worked. When examining the aircraft in the hangar, I found that damage had been done to the underside of at least twelve feet of the fuselage, severely wrinkling the skin and crushing the fuselage structure. The pilot had been very, very lucky. It was lucky for us, too, that the aircraft was still on its wheels as we could park it outside to wait for the Queen Mary to take it back to Fairey's.

7 Queen Mary was the common name for a long low articulated road vehicle used for transporting aircraft, usually partly dismantled, or the remains of a crash.

The large new Nissen hut gym in "B" Camp became the social centre for the station. By now we had our own chaplain so there were regular Sunday morning church parades. There was a cinema show once a week and there were occasional ENSA visits. Dances were held weekly and were very well supported. The station dance band left a little to be desired but made the evenings enjoyable. Tins of talcum powder were used as a lubricant on the bare concrete floor and we all left at the end of the evenings with our trouser legs white with the stuff. Winter evenings were long that far north, and another worthwhile entertainment was listening to classical music on the Wren officers' gramophone in their small wardroom. As the evenings lengthened it was possible to re-institute inter-departmental hockey matches. I don't think I played much rugger, or cricket in the summer, but I kept myself fit enough to enter several events in the summer athletic sports. A handicap to any of these outdoor activities was that there was no sports field at "B" Camp so all games were at "A" Camp. By my time there was no direct road between the camps so one had to go round by the public road, on one's bike, a distance of nearly a mile and three-quarters, having already changed into one's games strip. The 1944 athletics day was run in beautiful summer weather. I probably entered in more than one event but the only evidence I have is a photo of me winning the half-mile race, Cdr Wailes being the finishing line judge.

In the spring of that year Jimmy Barrett suggested that I might get a J2 MG Midget for rebuild on the station. I cleared this with Cdr (E) and I took over an unused Nissen hut in the technical area. The purchase was done in circumstances quite different to the comparative peace and calm of life at Fearn. My summer leave was due in June and I chose to take it at the end of the month, which turned out to be three weeks after D-day. A round trip to London and Edinburgh allowed me ten days at home.

The night sleeper train from Inverness to Euston on 29th June ran as usual but was severely delayed at Watford the following morning, an ominous sign of what was to come. For the first time I heard the crackle of the "doodle-bug" unmanned V1 flying bomb, the pause after the engine had stopped, then the massive explosion of the 1,870 lb bomb. After an undisturbed breakfast in the Euston hotel restaurant I was ready to venture forth into London's streets. June 30th had dawned and I was in the middle of the V1 bombardment of London.

My morning call was to the offices of *The Aeroplane* and *Aeroplane Spotter* magazines to renew my acquaintance with C. F. Andrews whom I had met two years before on my way to Henlow. They shared an upper floor in the Temple Press building in Bowling Green Lane, EC1, from where large windows gave a panoramic view over the roofs of neighbouring buildings. We were surrounded by the sight and sound of the non-stop barrage of "doodle-bugs". By that time in their flight they were down to only several hundred feet of altitude so were easily visible. Explosions were heard every few minutes, many producing highly visible clouds of smoke and debris. I sensed that *The Aeroplane* employees were on edge – indeed frightened, and I don't blame them for that because I was, too.

Mr. Andrews invited me to lunch with him at his club in St. James', taking a bus almost door to door. We chose to sit on the upper deck and on the front seat where the window had not been screened against flying objects. We could see everything as we progressed down Shaftesbury Avenue towards Piccadilly Circus and Piccadilly. There was nothing between us and the Circus when, suddenly, a cloud of smoke and debris shot out of the street leading to the Regent Palace Hotel, accompanied by an enormous explosion. The bus came to a standstill and rocked on its wheels as glass cascaded from the windows in the buildings all around us. People could be seen staggering drunkenly about the pavement and falling to the ground. In a moment the street was a sea of glass. This was Friday, 30th June 1944 at 12.30 p.m. when a V1 fell on the top floor of the Regent Palace Hotel – headlines on the national papers the following morning. My host and I were both surprised that the bus was undamaged. Within a minute the driver set off again, crunching over the glass to Piccadilly Circus and along Piccadilly. We consumed several large gins in the club. I noticed that we were both shaking with shock; I had experienced the Home Front under conditions of war but thanked the Almighty that this experience was but a blink in the awfulness that Londoners were put through.

I had booked into an officers' club on Piccadilly for the night. It faced Green Park and I was allocated a bed a few feet from the front window of an upstairs room, overlooking the park. The snarl of the pulse-jet engine of the V1s, followed by the inevitable explosion, went on all night. It was unnerving!

The following morning I was relieved to be able to get out of London by train to Watford to see the MG J2 sports car I had been negotiating over for several weeks. After a short drive under trade plates to confirm that it was roadworthy, I parted with about £120. War had not stopped all peacetime railway services and I was able to book a van to take the car all the way to Fearn station where it was delivered a few days after I returned from my leave.

By this time I had resolved to get away from London by the night through train to Edinburgh as soon as possible. After one or two more calls, in the evening I set out for Kings Cross. Underground trains were in chaos due to the V1 bombardment and I reached the main line platform to see the rear lights of the Edinburgh train disappearing northwards. My last hope of escape was to take the last train of the day to the north which terminated at Newcastle-on-Tyne. I took it and was deposited in the north of England, in one of the Victorian era's most elegant but draughty railway stations, at an early hour of a Sunday morning when no waiting rooms were unlocked. I spent an uncomfortable period of some hours on a wooden platform bench, eventually boarding the one Sunday train – and an upholstered seat - to Edinburgh. The time of my arrival at home was unplanned but the welcome was warm and I enjoyed a late lunch. My final return to Fearn and the Highlands was almost like going on holiday.

Jimmy took me over to Fearn station to collect the MG and somehow we managed to get it started and I drove it back to the base. We spent many evenings down on the airfield getting our hands dirty and made some progress. Inevitably my posting away in October found the re-build unfinished, but with much goodwill at Fearn arrangements were made to get the car to Edinburgh and it ended up in a garage at my father's department at Edinburgh University. This was not very satisfactory and the story ended when the car was sold to the Edinburgh MG agent, Scottish Motor Traction, while I was still in Australia. I never drove it again!

Langrishe winning the half mile race, Cdr Wailes on the tape

During these summer evenings on the airfield we stumbled on what we were told was a regular feature in the air. An unusual aircraft engine sound attracted our attention one evening. My Aircraft Recognition mode immediately roused me and, to my astonishment, we saw a Junkers Ju88 fly straight past the airfield from north to south. It was no more than a few hundred feet up and travelling at a very modest speed. It was a few years later, when I was corresponding with an aviation historian in Inverness, that I heard the story. Apparently, this intrusion was well known at the time and the Observer Corps had given it a nickname. It was a regular flight, originating in Norway. The Ju88 made landfall over the Orkneys, looked at all the airfields on the islands, then turned south, overflew the three airfields around Wick, over Tain, Fearn, Alness, Evanton and Inverness, there turning east to cover the row of airfields along the coast to Fraserburgh, of which there were at least eleven. It does seem a bit odd that no effort was made to intercept the flight at least once – or perhaps there was, but nothing was said.

Our routine of major inspections and minor repairs carried us through the summer, but there was not a great deal of work to occupy the young AEOs of any department. We did upset Cdr(E) one lovely summer day. I was trundling back to ARS driving the Coles crane, the Electrical "subbie", S/Lt(L) Smith RNVR, had collected the Fordson tractor and was approaching the hangars, the AEO of ERS, S/Lt(A/E) J. S. D. Tudor Owen emerged from his hangar driving the ride-on Lister platform truck – when Cdr(E) emerged from his hut! He was almost speechless to see his young officers behaving like schoolboys. Driving things was supposed to be a rating's job. I expect Jimmy Barrett had difficulty hiding his amusement; he would understand. Actually, the Fordson tractor was a difficult machine to drive. It was a general-purpose version of the agricultural tractor, having tyres on the wheels instead of all-metal wheels. The three-speed gearbox had straight-cut gears and no synchro-mesh, so the normal procedure was to start from a standstill in the gear you intended to use. One of our men found how to change gear on the move, otherwise this was impossible without making a really terrible crashing noise.

Far in the north as we were, and with everyone recognising the importance of keeping secrets, it was a complete surprise to hear that the invasion of northern France had begun on 6th June 1944. In our immediate area we were aware of training activities. Early in 1944 a stretch of beach and the land immediately behind it, a mile or so to the north of the airfield, and extending for another mile, were made "out of bounds" to everybody while amphibious landing practice was carried out under live fire. It seemed rather hazardous but there were no reports of neighbouring civilians being killed. Within a few weeks all the barbed wire was cleared, and the land was reopened. After D-day several aircraft appeared from the south with "invasion stripes", wide white bands on both wings, top and bottom, fore and aft.

Consolidated Liberator FL927/G. This aircraft was from a series of Mk.III bombers built on the B24D line. It was extensively modified by Scottish Aviation at Prestwick, Ayrshire, to become the prototype Liberator GR.V. Equipment included ASV Mk.II, Leigh light in housing under starboard wing (to assist night-time attacks on U-boats), rocket projectile sponsons. As depicted this aircraft also carried a mock-up AS9 radome under the nose. It was this mark of Liberator which was flown by the two Coastal Command squadrons which were based at Tain from Summer 1944.

In July and August Tain became a front-line Coastal Command base when Nos. 86 and 311 (Czech) squadrons moved in with their Liberator Vs, their job being anti-submarine patrols along the Norwegian coast and out into the North Atlantic. On several occasions we saw aircraft returning with damage which we were told was the result of aggressive gunnery by the U-boats. The Germans had appreciated that the last few seconds before dropping bombs or depth charges necessitated the Liberators and Wellingtons flying a straight approach. Instead of diving, the U-boats manned all their deck guns and fired on the aircraft. Many were shot down or badly damaged. Also, we had heard with excitement reports of Operation *Tungsten*, the raid on the *Tirpitz* during the spring of 1944, carried out by the largest aircraft carrier task force yet deployed by the Royal Navy. It was a baptism of fire for the Barracuda, flying from *Furious* and *Victorious.*

Living in our own little world it was important that everybody should be kept happy. One way was through good food. North of Inverness rationing of fresh meat and eggs was suspended and there were plentiful supplies of dairy products, poultry and fish. Within the financial constraints of Cdr (S)'s daily allowances we were fed well in the wardroom and I believe it was the same in the ratings' messes. It was commonplace to see personnel on their way to Fearn station to start their leave carrying a parcel of unrationed food, often a fresh chicken or a fish from Balintore, to take home. Christmas dinner in the "B" Camp wardroom was a splendid affair. We tended to keep to the Station for our entertainment, but I did make an effort to celebrate my 21st birthday in January 1944 by taking a Tilly-load of my friends to the Royal Hotel in Tain for a very enjoyable dinner.

I took my Christmas leave after Christmas and went home to Edinburgh to welcome my brother John back, after eighteen months as a Prisoner of War in northern Italy. I have mentioned that we'd had a Red Cross card in July 1942 giving us the bare details. He was a Lieutenant, Royal Artillery, and on 6th June 1942 had been captured with his battery in the desert behind Gazala by a German 88mm battery after they had run out of ammunition. Luckily, he was taken by the Italian Air Force to the Po valley, where he was in two different camps consecutively. At the armistice with Italy in September 1943, the Italian camp commandant at Fontanellato opened the camp wire and the Allied troops marched out in columns of three, dispersing into the surrounding countryside. Some men made for Switzerland, but John was determined to cross the Allied lines to the south. He kept to the mountains all the way, walking some four hundred miles to the River Sangro which he and his companions waded across at night after passing through a German artillery battery. This was in November 1943 and, after being passed fit in a military hospital, he was taken across the Mediterranean to Morocco where he took a lift in a USAAF C-54 transport aircraft to Prestwick airport in Scotland. Army transport took him home, where he arrived on Christmas Eve. My parents had not heard from him for several weeks, so his arrival was a great and happy surprise. John was Mentioned in Dispatches for his epic walk.

It must have been about this time that we experienced almost a revolution in our mode of dress. It had become obvious to all us young FAA officers that the traditional 'No.5' reefer jacket was particularly unwearable in the context of aeroplanes. At that time even aircrew would fly in their uniform jacket with a "Mae West" survival jacket on top. Special air suits or overalls were available but not obligatory. It was equally inappropriate for AEOs to clamber about aeroplanes and to get oil or other fluids on a tidy suit. So the basic design of the army battledress was adopted, but made of a dark blue serge. It could be worn by all officers who showed their rank on epaulettes which comprised a curtailed rank ring with the standard 'curl'. A standard cap was worn – no berets for some years. Ratings continued to wear their age-old uniforms but technicians wore overalls for their work. I found the new uniform most agreeable. As we were to find out only a few months later, the change to tropical khaki in warmer climes was even more liberating

The last three months of my time at Fearn were taken up with the periodical and random work provided by the two new squadrons, Nos. 714 and 717. It was at about the time that I was informed of my next appointment that these two squadrons were preparing for their transfer to the new Royal Naval Air Station at Rattray, south-east of Fraserburgh (also known as Crimond). My new appointment came through in October; I was to report immediately to RNAS Ludham, Norfolk, *HMS Flycatcher*, to join MONAB1 "for Air Engineering duties". Nobody at Fearn knew what a MONAB was nor where it might be deployed to, or for. The move was not too much of a logistics nightmare because my home was on the way south and the railway system I would use was down the east coast. There were no instructions about uniforms for a foreign station so all I needed to do was to sort out all my kit at home and be off to Norfolk. It was a beautiful morning when I set off from Fearn on 17th October 1944 – with regrets, I must say, but with good wishes from Cdr (E) and all my friends ringing in my ears. Cdr (E)'s last words were that he had recommended me for promotion to Lieutenant. The country was starting to change to autumn and, seen from the train, the leaves of the birch trees in the Strathspey were a beautiful golden colour.

Now for the next adventure!

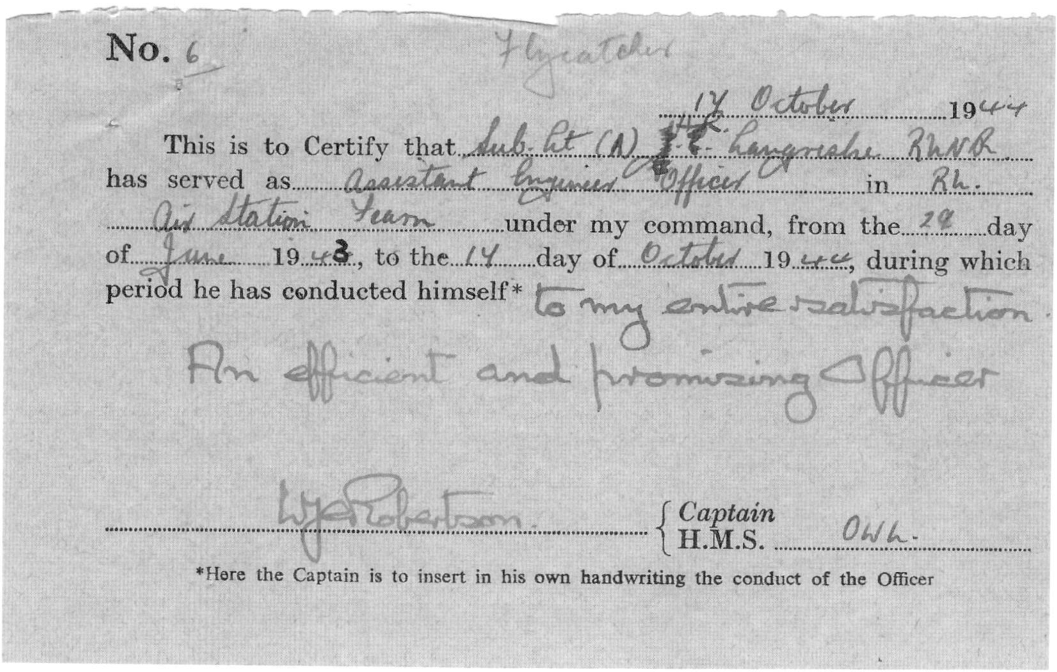

No. 6 Flycatcher

17 October 1944

This is to Certify that Sub. Lt (A) [illegible] Langrishe RNVR has served as Assistant Engineer Officer in RN. Air Station Fearn under my command, from the 29 day of June 1943, to the 17 day of October 1944, during which period he has conducted himself* to my entire satisfaction. An efficient and promising Officer

W J Robertson. Captain H.M.S. OWL.

*Here the Captain is to insert in his own handwriting the conduct of the Officer

Flimsy from RNAS Fearn 7.10.1944, on my appointment to RNAS Ludham, HMS Flycatcher. Note this is from Capt. Robertson who took over from Capt. de Courcy Ireland in February 1944. The intervening months had been interesting and productive for me.

Postscript

RNAS Fearn - "A" Camp wardroom photo - taken in 1986

RNAS Fearn – Remains of hut housing the first Air Engineers' offices and garage 1943. (1986).

My wife Pamela and I visited Fearn in the summer of 1986. We obtained permission from the farmer who then owned the land on which the airfield had been built to drive round and see my old haunts. All the wartime hangars had gone although one or two new sheds had been built The control tower was derelict but still standing to its full height. The three runways could still be seen, in part intact or stripped of their hard surface. One or two Nissen huts had been retained. The south-east corner of the airfield was being dug for gravel and sand. The derelict remains of a number of huts in "A" Camp still remained, but the entirety of "B" Camp had been removed and much of it dumped on Balintore beach, leaving a grass field – which it had originally been. It was all a bit sad!

PART THREE: SUPPORTING THE BRITISH PACIFIC FLEET, 1944 – 1946

An adventure down under

Introduction to MONABs

I left RNAS Fearn, Easter Ross, on 17th October 1944 on appointment to RNAS Ludham, Norfolk, to join MONAB1 which was just forming there. MONAB stands for Mobile Operational Naval Air Base, a complete self-administered unit sent to Australia and the Far East from 1944 onwards. A total of 9 were deployed.

Ludham was in the north-east of the county, on the edge of the flat lands of the Norfolk Broads. Our Station was built as a RAF fighter airfield, loaned to the Navy as a base for the formation of a series of MONABs. Why we should have to form up on an airfield was not disclosed; at no time did we see a naval aircraft and dispersal of hundreds of people with virtually no motor transport made our lives very difficult. With two notable exceptions none of us knew anything about the organisation we had joined. An organisation known as the Mobile Naval Airfields Organisation (MNAO) had been working for a year to put the concept of mobility into reality and was commissioned as an independent headquarters unit at Ludham on 9th September 1944. The enormity of the idea became obvious to us as we gathered together at Ludham. MONABs Nos. 1, 2 and 3 were all assembling during October and November 1944, each with 500 to 600 personnel. If ours was typical, MONABs almost at once set up their own ship's organisation and administration; captain, commander, No.1 and the hierarchy of regulating office all doing their jobs as usual. As we had arrived in ones and twos from Naval establishments far and wide, there was no formal introduction or explanation of the MONAB idea and how it was to be implemented. Indeed, many of the non-FAA officers had never been near the Fleet Air Arm. What we learned was picked up in conversation with those who had arrived before us, the most important information being that we were destined for Australia, with the possibility of advancing up into the Pacific islands.

Most of us had just four weeks in which to get to know our colleagues and the task we faced individually. Not only that, we had to undergo inoculation and vaccination, then take a fortnight's embarkation leave during which we had to equip ourselves with tropical uniforms. I went home to Edinburgh for my leave, travelling during the night after having my inoculation and feeling just a little "rough". My father, who had been an Army doctor for twenty years, including for the whole of the Great War on the Western Front, was furious that I should have been sent home so soon after inoculation. I thought it significant that when he said good-bye to me in the dark at Waverley station, his demeanour suggested that he considered my departure as the last time he would ever see me. My brother said the same thing had occurred to him when he set off to rejoin his RA regiment for the invasion of northern France in June 1944. The MONAB working uniform was standard army khaki serge battledress with boots and gaiters, displaying "ROYAL NAVY" shoulder flashes on the sleeves. Our "Blues" uniform was packed away until we were ashore in Australia in winter rig. During the summer months we were to wear tropical khaki, shirts and shorts during the day and a trouser suit with long sleeves in the evening or when off duty ashore. We were told to bring a full set of tropical whites, but these were never worn ashore. I never went aboard one of HM ships at a time when I should wear white, so the uniform was never used. I was able to equip myself with all that was required at the Edinburgh branch of Gieves.

Ludham was a typical wartime RAF airfield. Domestic sites were dispersed well outside the airfield itself and could accommodate 190 officers and 1,519 ratings in dormitory huts. I have no recollection of wardroom facilities. Because our motor transport was not on site we had to walk or march everywhere. Our technical department was allocated one Jeep and one two-wheeled trailer from the Station MT section which, I seem to remember, was driven by all and sundry.

For many reasons, not least logistical, all the MONAB equipment was assembled in the north-west of England, a convenience for proximity to Liverpool docks. This was to be the base from which a monthly stream of ships left for Australia. The Ministry of War Transport had brought together an impressive fleet of ocean liners, all of which had already served as "troopers" to and from the Middle East and India and across the Atlantic. For this purpose, many of the freight holds were fitted out with double- or triple-level bunks. Other holds were used for personal and official luggage for the several thousand passengers. Motor vehicles and stores, all the equipment for accommodation, administration, cookhouse and sick bay, collapsible hangars, airfield and air engineering equipment and stores, some of which was packed into its own dedicated wheeled vehicles, all of this was loaded on to a merchant ship which sailed in mid-December for Sydney.

Having no equipment at Ludham, and no Air Publications, there was nothing constructive we could do except to get to know each other, particularly our own ratings. The senior officer of our technical department was Lt Cdr (A) (AE) G.A.G. Morris RNVR, a pre-war student of the De Havilland

Grumman Avenger II JZ296, 385/W of 857 Sqn. flying over his squadron's ship, HMS Indomitable, somewhere in the Pacific, 1945.

Aeronautical Technical School at Hatfield. He had qualified as a Licensed Aircraft Engineer and had worked for Imperial Airways until volunteering for the Fleet Air Arm. He was one of the two officers I have referred to who had been key figures in the MNAO. He and Cdr (S) E.H. Taylor RNVR, our Cdr (S), a pre-war reserve officer in the London Division, had both been responsible for working on the schedules of equipment for MONABs. Both men were personalities who got on well with everybody, officers and ratings, and were crucial in creating an efficient unit.

I was told I was to command a unit known as MS1, or Mobile Servicing (Unit) No. 1, the very first of a long list, of which two were assigned to each of the standard MONABs, Type A. Each MS unit was dedicated to one type of aircraft. Mine was for the Grumman Avenger torpedo-bomber, and my opposite number, S/Lt (A) (AE) J.W.W. Clemes RNVR, took MS2 which was for the Chance Vought Corsair single-seat fighter. Clemes was a graduate of the first year-long course for RN cadet air engineers at the Loughborough Technical College. Also in MONAB1 was a unit known as MM1, a maintenance unit which had similar duties to an ARS and which was a responsibility of Lt Cdr Morris, assisted by two sub-lieutenants (AE) and one air electrical sub-lieutenant. It was equipped and stored for Avenger, Corsair and Martinet target-tugs. The addition of the latter was because the mother airfield was expected to house a second-line squadron devoted to Fleet Requirements. This became 723 Sqn., of which more later. Post-war assessments of the problems of finding enough men for the huge requirements of the British Pacific Fleet made it clear how serious and difficult the procedure was. It has always seemed to me that the continuing expansion of the Fleet Air Arm, by the commissioning of new Air Stations and new aircraft carriers and air squadrons, made it almost impossible to find enough senior and experienced technicians of all specialities and rates to man the completely new MONABs. My small unit absorbed two air artificers, two petty officers and three leading hands, all of different trades. The practical interpretation of the MONAB concept was not only to provide technical skills and equipment but also to be able to supply a range of spare parts for the specified aircraft types, sufficient for a certain period of weeks. MS units were each expected to support a squadron of up to twelve aircraft for six weeks before having to call on Naval Air Stores for replenishment. It did not work out according to plan - but that will emerge later in my story. What did concern us all was that none of us had worked on the Avenger, nor on its engine, the Wright R-1600 Double Cyclone, and none of my men had US-standard spanners in their UK-issue toolkits.

Chance Vought Corsair I and II, of JT series, of 738 Sqn. from USNAS Lewiston, Maine, where the squadron was engaged in advanced carrier training, 1944.

I am at a loss to describe our daily routine because there was very little we could do without even one technical publication to read. My most vivid memories are of the impact made on our perception of the war by the goings-on at our neighbouring airfield. Some five miles to the south-west was Rackheath, the eastern-most of a string of three USAAF heavy bomber bases of the 96th Combat Bomb Wing, equipped with the B-24 Liberator. The other two were Horsham St. Faith and Attlebridge; each base housed a Bomb Group of four Bomb Squadrons, each equipped with twelve aircraft. As we roused ourselves in the morning, we would hear the rumble of engines from Rackheath rolling across the flat Norfolk countryside as aircraft were checked and warmed up for the day's mission. That October and November we enjoyed a period of fine weather and on four or five days each week an impressive number of aircraft were dispatched from each of the three airfields. By that time the USAAF had developed a defensive box formation for their heavy bombers, to give the air gunners a good field of fire to protect themselves and their formation neighbours. Each squadron formed its own box as one by one the aircraft took off, then the squadrons slowly drew together to form the Bomb Group box, circling and climbing together. They were marshalled by a controller in an "Assembly Ship", a dead-beat Liberator which was painted overall in a garish but very visible colour scheme. The controller aboard signalled the formation by a shower of coloured Very lights as they slowly climbed. One particular day when there was evidently to be a "maximum effort" Rackheath put up a formation of around forty aircraft. At the point when the box finally turned towards the coast, I could see the Horsham Group approaching and, looking again, I could see the Attlebridge Group behind them, making a total of well over one hundred B-24s. It was one of the most exhilarating and awe-inspiring aviation sights I had ever seen. But that day the excitement was not yet over; several hours after the Liberators had disappeared, the air was once more full of noise and we rushed outside to see a huge formation of over a hundred P-51 Mustangs fly over, evidently planning to catch up with the Liberators as they entered the German fighter zone over Holland. The end of the day saw things in rather a different light. The Rackheath formation had broken up on its flight home and many aircraft were visibly very badly damaged – engines labouring, stopped or even missing, chunks of airframe shot away, undercarriages broken and hanging down. Aircraft with dead or wounded aboard, or so badly damaged that they could not land on a runway, signalled to their control tower with a shower of red lights and they were given priority to land on grass away from the duty runway. We watched this performance with awe – and, I might say, with the greatest admiration.

Consolidated Liberator B-24J-CF No. 44-10560. Aircraft of this mark were built in huge numbers and would have been used by the squadrons at Rackheath

Journey to Australia

So, by the 18th November we were almost ready to go. We had been formally commissioned on 18th October as *HMS Nabbington*, under Cdr G.A. Nunneley RN, but there was one final thing to do, to identify our luggage. We were told to paint or apply our name and the characters "MONAB1 SYDNEY" on all our pieces of kit. We were told that our train journey to Liverpool would be by night, following a very tortuous route across the midlands. We were taken by lorry to Wroxham station – the whole of MONAB1 together with its luggage and a sandwich supper – and the special train finally set off for Liverpool sometime around 23.00. Progress was slow and at times zero. It was light the following morning when we stopped at a complex of railway sidings on the Birkenhead side of the Mersey. I put my head out of the carriage window and was surprised to see that we were one of a string of trains waiting to unload. From the middle of nowhere we were taken by bus to an otherwise deserted camp where more sandwiches were produced. After several hours there we once more boarded a fleet of buses and were driven to Gladstone Dock, Liverpool. We were to board what at the time was a very large ocean liner, *Empress of Scotland*, of the Canadian Pacific Line. She had originally been named *Empress of Japan* but for obvious reasons this name was considered unsuitable for an Allied troopship and was changed in 1942. The ship had a displacement of 30,575 tons and had a speed of 22 knots (builders' figures). She had been trooping since 1942 and had arrived at Liverpool from South Africa only a few days before starting the Sydney run. As MONAB1 luggage was unloaded from lorries on to the quayside there was consternation when it was seen that all pieces were marked with our destination, in very large letters! Nobody was supposed to know! I must say that the organisation for settling everyone into their various cabins and troop decks went remarkably smoothly – we were told that there were to be over three thousand people aboard. The junior officers from MONAB1 were put together into several upper-deck "state" suites, the numbers in our cabin being made up to twelve by several Air Branch people I had met on courses and who were to take staff jobs in Sydney. They included my old

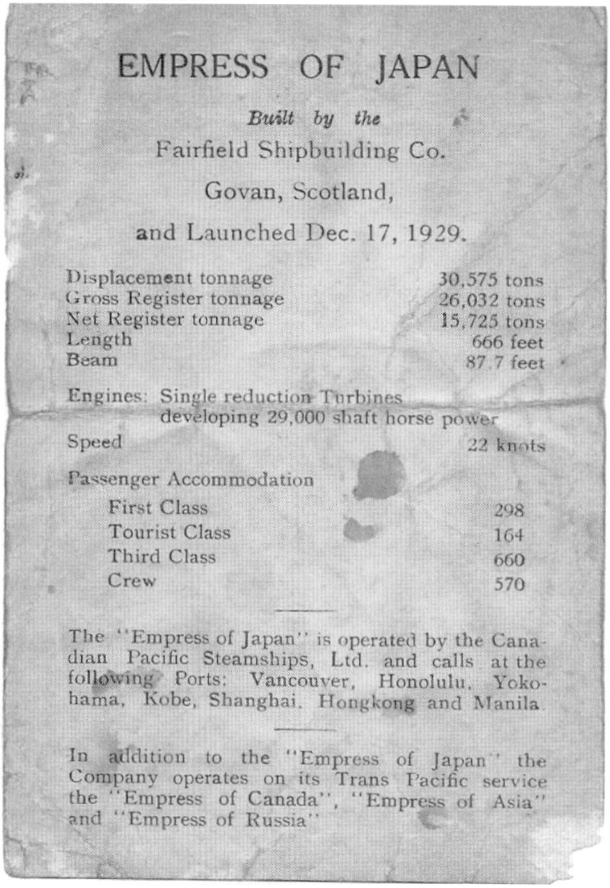

EMPRESS OF JAPAN

Built by the

Fairfield Shipbuilding Co.

Govan, Scotland,

and Launched Dec. 17, 1929.

Displacement tonnage	30,575 tons
Gross Register tonnage	26,032 tons
Net Register tonnage	15,725 tons
Length	666 feet
Beam	87.7 feet

Engines: Single reduction Turbines developing 29,000 shaft horse power

Speed 22 knots

Passenger Accommodation

First Class	298
Tourist Class	164
Third Class	660
Crew	570

The "Empress of Japan" is operated by the Canadian Pacific Steamships, Ltd. and calls at the following Ports: Vancouver, Honolulu, Yokohama, Kobe, Shanghai, Hongkong and Manila.

In addition to the "Empress of Japan" the Company operates on its Trans Pacific service the "Empress of Canada", "Empress of Asia" and "Empress of Russia"

Particulars of the Canadian Pacific ocean liner Empress of Japan when built. Her name was changed to Empress of Scotland after the outbreak of the Pacific war against Japan in December 1941.

friend "Robbie" Robertson who, because of his age, had already put up his second ring. The cabin had been stripped of its furniture and equipped with twelve bunks in double-deck formation. We had two portholes to the outside of the ship. With a considerable amount of personal luggage there was very little space to move around. We had one built-in wash basin which was augmented by separate wash, shower and WC facilities. For messing we were allocated to one of the "cabin class" restaurants towards the stern of the ship. The officer passengers were allocated the first-class lounge as their social space and the senior officers used the old first-class restaurant. Our ratings were all together in troop decks well below the waterline, created out of the cargo holds, with no access to fresh air. This was quite uncomfortable during the early part of the voyage across the north Atlantic when the ship encountered rough weather – in fact, the storm we passed through compelled the captain to bring the ship almost to a standstill during the night and turn into wind. The following morning, I went up on deck and was surprised to see that the ship's bow frequently dipped into the sea.

So, coded as "J1", the first troopship on the regular run to Australia, *Empress of Scotland* departed Liverpool on 20th November 1944, bound for the Panama Canal.

MONAB on Active Service

Life on a troopship was by no means unbearable for the officer passengers. If the accommodation was cramped, at least it was above the waterline, and the officers were very well provided for in the first class lounge – except for the fact that the ship was "dry". For fresh air we had the run of the boat deck, either for exercise or just for lounging in the sun. Junior officers were designated "Troop-deck Officers", which necessitated visiting one's troop deck from time to time, an unpleasant chore. However, this duty also required a presence at daily boat drill when the denizens of the troop decks were compelled to muster in the light of day and be counted, a procedure which kept us standing for a good hour. As we approached the tropics this became a rather more pleasant use of our time than it had been at the start. It was also an opportunity to couple names with faces. During this procedure I had noticed a middle-aged-to-elderly gentleman watching us from outside his cabin on the deck above. After a couple of days, he motioned to me to come up after I had dismissed my boat drill party. He was the ship's surgeon, one of the liner's crew. An Irishman, and graduate of Trinity College, Dublin, where my father had done his medicine forty years before, he was looking forward to retirement from the Canadian Pacific line, with whom he had served for many years. Rather than going to the ship's wardroom for his morning coffee, he chose to invite guests to his day cabin, and I was lucky to be one of them on many mornings for most of the following four weeks. It was a light and airy cabin, furnished like a country cottage, with chintz covers on the chairs. It was very enjoyable to drink a cup of coffee in such civilised surroundings. There was no such thing as an electric kettle, so water was boiled in the surgeon's instrument steriliser! After a few days an army officer appeared for coffee, somebody I immediately recognised as being a senior at my prep school in the 30s. He was Capt. J.D.B.Richardson of the King's Dragoon Guards. He was travelling to take up his appointment as ADC (aide-de-camp) to the new Governor General of Australia, the Duke of Gloucester, who was due to take up his post early in 1945. I must say that this morning ritual was a bonus which I coveted and which I was privileged to enjoy entirely separate from my fellow officers.

MONAB1 was enabled to travel as a self-administered unit – probably to the delight of OC Troops, the officer in charge of all uniformed passengers in a troopship. We had our own suite of offices for Captain, Commander, No.1, Master at Arms and the Regulating Office. We operated our own rota of Officer of the Day and, although I cannot now remember any details, we probably had Captain's and Commander's Requestmen and Defaulters. Our initial strength was thirty-one officers and over five hundred men. As we were the first troopship to Australia there were many officers from commander downwards who were travelling individually to open up and run various staff and administrative offices in Sydney, and to set up Naval Barracks, Naval and Air Stores, Fleet Pay Office and Fleet Mail Office. Sydney was to be the operating base of the British Pacific Fleet. The passenger list also included a few women in uniform. Some were WRNS officers, but the majority were nursing sisters of the Queen Alexandra's Royal Naval Nursing Service who were to staff a new RN hospital. Catering on the ship could not be faulted; the ship had stored in Cape Town early in November in preparation for her next long run to Sydney. The food was every bit as good as we could remember of the times of peace at home.

We left England under leaden November skies, the week the Ministry of Food allowed ice cream to be restored to our menus. Classified as a fast ship we were only escorted for about forty-eight hours, then proceeded unaccompanied, zig-zagging all the way to the Caribbean. It was quite a shock to experience the searing heat of the Caribbean as we approached Panama. However, we changed into khaki tropical uniforms and put our serge battledress away, not to be used for several months, and felt better! We passed through the breakwater at Cristobel at 09.00 and were eventually alongside in mid-morning. It had taken us about eleven days from Liverpool. Most of the day was spent oiling and re-storing foodstuffs. During the afternoon we were allowed ashore to exercise in the docks and in the evening we were entertained to two separate live shows in the dockside warehouse. The excruciating discomfort of sitting on the hard, cobbled floor was passably mitigated by whatever item of clothing we could find to use as a cushion. Passage of the canal started at about 07.00 the following morning and took some nine hours. The spectacle kept most passengers on deck until chased in by the heat or the call of lunch. The canal and a strip of land on either side were United States' sovereign land, consequently the canal was under the control and management of the United States. Every ship was in the hands of a Canal pilot and, during the war, armed US soldiers were distributed about the open decks. Pilot and soldiers were taken off by tender after the ship had finally entered the Pacific Ocean.

We were very lucky with the weather. Day after day the Pacific lived up to its name and we thundered along in predominantly calm seas and glorious sunshine. As we got to know each other and explore each other's talents we discovered that we had several musicians amongst us, including a very competent pianist and one person who was an experienced writer of light musical comedy. It was decided that MONAB1 would put on an evening entertainment in the officers' lounge. In spite of the very limited time and facilities we staged a highly professional performance! Of course, many lyrics and poems poked fun at the Royal Navy and our circumstances. We even put a chorus line on stage to sing and dance, all performed with gusto and panache – but questionable skills. My notes record a "final concert" during the evening we entered Sydney harbour, so we must have impressed our audience enough to encourage us to expose ourselves one last time.

Some might find an ocean passage to be one without interest outside the ship. I had been interested in birds since childhood and I was fascinated to see so many species on the wing. I knew that most seabirds of the northern hemisphere spent many months of the year away from land and these we saw in the North Atlantic, thousands of miles from land. In the Caribbean, quite suddenly, there was a change to tropical and sub-tropical species such as pelicans and man-of-war hawks (frigate birds), the latter perpetually on the wing during daylight. As we worked our way south-westwards across the Pacific, albatross appeared. They seemed to be using our ship as a means of propulsion; they would join the air wake astern of the ship, glide forward alongside until level with the bow when they would curve away aft to rejoin the wake and do it all over again – and not a movement of the wings at all.

I have two more observations to make, one being the flying fish, seen in tropical waters, which seemed to "fly" an astonishing distance a foot or so above the sea. The other, which in modern times would be considered outrageous, was the daily disposal of "gash" through a chute in the side of the ship. It formed an unbroken stream into the distance. In mitigation, at that time there was no plastic packaging material so most dry foods would have been paper-wrapped, a material which was readily bio-degradable. Nevertheless, as this was the standard way of waste-disposal at sea, such waste was contributing to the pollution of the oceans all around the world and had been for centuries.

Adding to our excitement as we approached Australia were the increasingly dramatic sunsets. Our meteorologist told us that the cause was a massive dust cloud which was being swept into the atmosphere from the farmlands to the west of the Great Dividing Range. We were to hear that Australia was in the middle of a severe drought which had already lasted several years and was bringing many farmers nearly to ruin.

There is nothing more dramatic than the final approach to Sydney directly from the east. We were told that the navigator, confused by the dust in the air, was having difficulty picking up coastal markers such as lighthouses, and our speed was reduced as night fell until the lights of Sydney Heads appeared in the gloaming. We were overflown by a Beaufort on coastal watch; it is doubtful that the radio officer on *Empress of Scotland* had the appropriate crystals to communicate with the aircraft by VHF – if indeed the ship had VHF at that time, but there was no subsequent action against us so we must have been accepted as "friend". Soon we began to see ordinary lights of the city and its suburbs twinkling low on the horizon, until, closer still, the horizon, some thirty to forty miles from north to south, was an unbroken string of lights. The sense of excitement was almost palpable as we slid slowly between the Heads and through the anti-submarine boom. Entry to Port Jackson, the body of water which includes the Harbour, is not straightforward. It requires an immediate turn to port (to the south) followed by a similar turn to starboard, when the full extent of Sydney Harbour becomes clear. We were then surrounded by myriads of lights on the sloping ground of the city which rises immediately from the waterside. Not only were there streetlights and house lights but many were moving – cars, buses, ships, ferries, trams and trains – it was a fairyland which five years of war had denied us at home. Once through the boom, all deck lights were switched on, to contribute to the wonderful occasion. It was magical and breath-taking. Our progress up the harbour continued while we went below for our dinner, and when we once more went up on to the boat deck, the ship had come to a standstill and was at anchor. The following morning, under a cloud of dust which seriously obscured the view of the harbour bridge, we slid slowly up the harbour, to come alongside at Woolloomooloo. The battleship *HMS Howe* was on the other side of the jetty. It was the Sunday before Christmas, 21st December 1944.

Setting up base camp at Nowra

We must already have known where we were to set up our base; it was to be at the Royal Australian Air Force (RAAF) Station, Nowra, New South Wales, near the coast approximately 100 miles south of Sydney. But what was to be done with some 500 men only four days before Christmas? There was no holding camp in the Sydney area although it was intended to set up temporary barracks at the Warwick Farm racecourse in the western suburbs of the city. We were kept on board *Empress of Scotland* for another night – the only passengers left on board and found ourselves used to tidy the ship and collect all the "gash". It was interesting to see the troop decks with the hatch covers removed and to ponder on how uncomfortable they must have been.

Although planning began immediately to move us down to Nowra, the simplest thing to do was to send everybody on a short leave period. Hospitality was in the very competent hands of the good ladies of Sydney and everybody found somewhere to enjoy Christmas. However, it was decided that a small advance party should proceed at once to Nowra, led by our Cdr (S), Eric Taylor. We spent one night under mosquito nets in makeshift accommodation at Warwick Farm, then the following morning some six of us set off by train for Bomaderry, the terminal station for the line to Nowra, on the north side of the Shoalhaven river. We had the utmost cooperation from the RAAF who picked us up, took us out to the airfield and for our short stay with them were extremely hospitable. Nowra had been a fully operational air base for several years, finally housing the OTU for Beaufort torpedo-bomber and reconnaissance squadrons. The unit had already moved out, leaving one aircraft awaiting spare parts, and a number of maintenance personnel who were also clearing the airfield of their equipment. They also continued to man the cookhouse. Eric Taylor was particularly busy getting to know the ropes of local suppliers and arranging local banking. We were there over Christmas so were roped in to help serve Christmas lunch in the airmen's mess. We were horrified to see how much uneaten food was returned to the cookhouse – there was no food rationing in Australia during the war. In the evening we were all entertained by the bank manager in Nowra, a very happy occasion. I suppose my reason for volunteering for this advance party was mainly curiosity; there was nothing concerning aircraft maintenance that I could do that would be likely to make our jobs any easier until we knew what tools and equipment were on their way. Regarding the airfield itself, it was a wartime-build to accommodate a rapidly expanding RAAF. It had two runways, both well over 2,000 yards long, and substantial hard standing areas – which still proved to be inadequate when the fleet arrived. It was on the edge of the eucalyptus bush, on a slightly sloping plateau several hundred feet above the level of the coastal plain. Views over partially cleared land to the coast were dramatic. Although the airfield was wire-fenced, it was necessary to check regularly that no kangaroos had strayed from the bush to the west on to the flying field. We were told that at times the RAAF people would go into the bush in the dark in Jeeps, taking small searchlights with them, to shoot the kangaroos as a deterrent. Apart from the north-eastern quarter down to the Shoalhaven, the surroundings were virgin bush, a type of country quite alien to us.

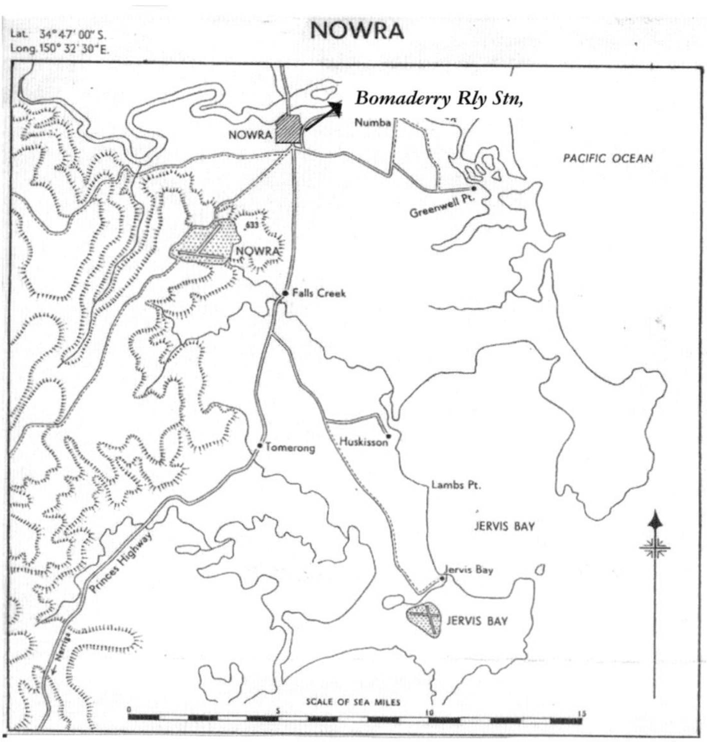

Map showing positions of Nowra and Jervis Bay airfields in relation to the town of Nowra.

The camp was on the north-east side of the airfield. Accommodation was all hutted – unlined wooden structures with corrugated iron roofs, built on wooden plank platforms 18 inches to 2 feet above ground level, presumably to allow air circulation to mitigate the summer heat. The officers' hutments and wardroom were at the north-east end of the camp, at the top of the slope from the airfield. The permanent residents such as the MONAB personnel all had single cabins in blocks of six or eight, with a generous verandah giving access. During my first stay I had a cabin in one of the huts at the top edge of the compound with a stunning view over Nowra to the north-east. It was almost like being on holiday! At night there was a continuous chorus from thousands and thousands of cicada beetles. To sleep under mosquito nets was obligatory at that time, even in Sydney; a wide variety of large and small insects was on the move in the dark. It is worth commenting that this closeness to "the bush" is an important aspect of the nature of Australia. It is not a country like ours, of neat fields, hedges and winding country lanes, of cosy villages around the village green with the ancient church in the background. Australia at that time still had the look and feel of a frontier town; Nowra's one main street was wide and the pavements were protected by a permanent canopy to keep the sun off shoppers. The only sealed road outside Nowra's main street was the Prince's Highway, the main coastal road between Sydney and Melbourne, which was a federal responsibility. Otherwise all minor roads were surfaced with sand and gravel, and those included the direct road from Nowra to the Air Station and the link from the Prince's Highway to the base. From time to time a grader would be used to iron out the ruts and corrugations. Another instance of this frontier atmosphere was the public telephone system. The exchanges were still entirely manually operated and most private houses were on shared lines. This frequently led to an operator helping a caller by disclosing that there would be no reply as "Mrs. So-and-So is out"!

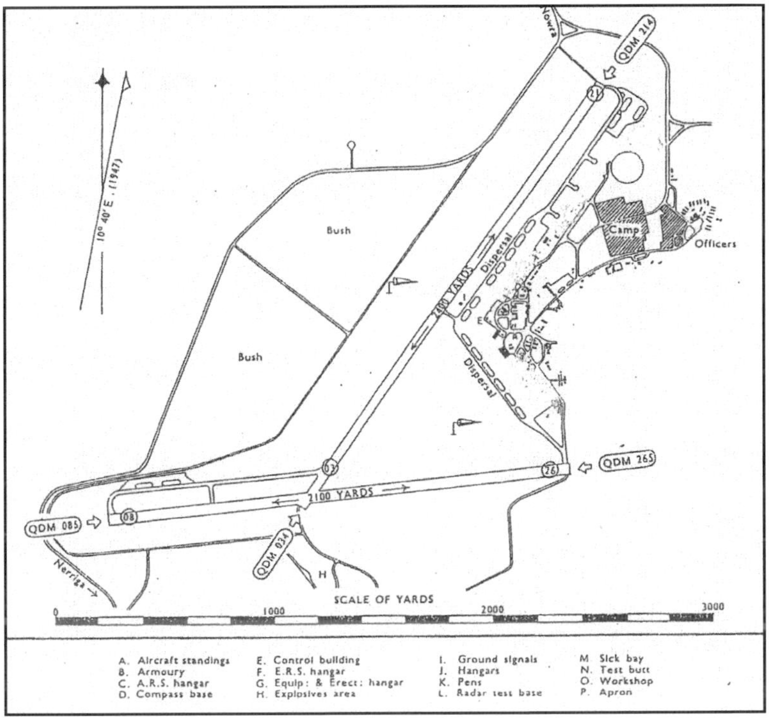

Plan of Nowra airfield as it was in 1944. Note that several items in the index at the foot of the plan did not exist in 1944-46. The large dispersal areas had not been finished at the time of the arrival of the squadrons in February 1945, necessitating the use of the runway edges for parking aircraft.

I had made contact in Sydney with a family to whom I was introduced by airmail by old friends of my family in England. I had explained to them what was going on and that I would be taking my Christmas leave at New Year. I was immediately invited to go and stay with them, a stay which started a friendship which only ended with the death of the youngest son several years ago. For a few more days our advance party continued to learn the ropes until it was decided that we were ready to go back to Sydney to enjoy a few days' leave. It was a surprise to find that a daily return passenger air service from Sydney (Mascot) to Nowra had already begun, and I was able to use it. This was to be the first instance of things happening before officialdom said they should. The aircraft was a Beech Expediter, carrying both RN serial number and Australian civil letters. The pilot was Lt Cdr H.A.P. Bullivant Royal Naval Reserve, the CO-designate of 723 Squadron; but 723 Squadron did not form until 28th February – as a Fleet Requirements Unit! It was 724 Squadron which had been planned as the communications squadron, but it did not form officially until 10th April. In spite of such outrageous flouting of official plans, everything worked! So, it was in a 723 Squadron Expediter, in the capable hands of Lt Cdr Bullivant, that I set off for Mascot on leave, to spend the New Year holiday in Sydney.

Beech Expediter II No. 549 of the RCAF. Nineteen of these six-seater light transports were shipped to Australia. They were issued to 723/724 squadrons and from December 1944 operated regular services around the RN air stations

My hosts were a GP doctor in an inner suburb of the city. The Loxton family had three sons and a daughter: the eldest son was a POW of the Japanese, having been captured in Singapore in the Australian army; the middle son was a career officer in the Royal Australian Navy (RAN) and had done a considerable amount of his early training with the Royal Navy early in the war; the youngest son, who became a good friend, had just left school and did national service as a sub-lieutenant in the RANVR (RAN Volunteer Reserve). I was introduced to the famous Sydney Cricket Ground and watched part of a New Year match between a New South Wales side and a "Selection" team. All were first-class cricketers, including the spin bowler W.J. O'Reilly and the Queensland opening batsman W.A. (Billy) Brown, both Test cricketers well-known in England, and there was some scintillating cricket, commented on by the usual noisy spectators on "The Hill". I was also shown a number of the magnificent beaches both north and south of the city. Of course, I was driven over Our Bridge, for which one of Mrs. Loxton's brothers was the supervising engineer.

Over the few days after the New Year MONAB1 personnel were transported by train down to Nowra, officially taking over RAAF Nowra on 2nd January 1945, assuming the name of RNAS Nowra, HMS

Nabbington, under the command of Cdr G.A. Nunneley RN. Quite why and how it had happened, the Navy was chronically short of motor transport. It is easy enough to say that the MONAB was transported by train. But approximately five hundred men had to be transported, with their personal baggage, from the mustering point at Warwick Farm to Sydney Central station to embark on a special train to Bomaderry, then to be taken the five or so miles out to the air station by bus and/or lorry. For the latter part of the journey, lorries were provided by the RAAF. There was now a requirement for transport to bring down the MONAB equipment and stores. This had been unloaded from the merchant ship which had brought it from England. Once more the need for motor transport was more than could be managed by MONAB transport itself; anyway, what had been supplied in the UK was below the numbers scheduled. The RAAF filled the gap. For many months this remained a sore point. Official records show that by the autumn of 1944, the MNAO had been re-tasking the Bases to lose a substantial part of their Mobile function, and that the first three or four would be expected to take over existing RAAF airfields on the Australian mainland. It may be that this influenced MNAO planning and execution of MT requirements. Certainly there was not enough of it for the normal operation of a Naval Air Station, let alone to carry the entire MONAB.

Bristol Beaufort VA, built in Australia for the RAAF torpedo-bomber squadrons and used extensively in the South Pacific, attacking the Japanese in the Solomon Islands and New Guinea. Aircraft of this type equipped the torpedo-bomber training unit at RAAF Nowra before the Station was taken over by the Royal Navy.

By the middle of January we knew what we had to work with. For the first time Clemes and I were coupled to our stores and equipment, and there were shocks, some good and some bad. We each had three vehicles, only one of which was self-propelled. This was a standard Bedford 3-ton 4x4 lorry, fitted with a strong work bench and vice and a battery charging unit powered by a small internal combustion engine, both these items being firmly fixed to the floor of the truck; we may also have had a compressor. The other two vehicles were 5-ton capacity four-wheeled trailer vans, each fitted out with special non-spill racking and stowage. One had a special corner furnished as an office, with non-spill shelves for books and catalogues. Each van had already been kitted out with the range of parts, accessories and equipment which had been scheduled back in the previous summer. What might be considered incomprehensible is that at no time did Clemes or I have to sign for anything, even the Bedford. Stretching things just a bit, Eric Taylor would say "but it's all consumable stuff, isn't it?" It would have made life much easier for airfield work if we had had a 15cwt. truck or a Jeep, but neither was available. The only vehicle which could tow one of the loaded trailers was an AEC Matador – which MONAB1 did not possess. One had to be summoned from Sydney and, I believe, was lent by the Australian Army.

Both the men of my unit and I were pleased to see that our equipment included the manufacturer's manuals for the Avenger and its Double Cyclone engine and their parts lists. They were all dubbed with Air Publication numbers but had obviously never been near the AP system in the UK. Being my own storekeeper I was worried by the fact that parts lists were in this non-standard mode and, although I have usually found it relatively easy to come to terms with new systems, this nearly defeated me – and the Naval Air Stores system. I recall that, if any, the "exploded" drawings we were accustomed to in British APs were not sufficiently explanatory. We were particularly pleased to see at least eighteen of each of several ranges of US standard spanners and screwdrivers – the latter being "cross-point", for a type of screw-head not yet adopted by the British aircraft industry. It was also our introduction to the socket spanner. This was the first time our teams had seen American tools, none of which were included in the standard Naval air toolkit. It was important that our men had them in their tool boxes, and any other special spanners needed for accessing nuts and bolts in otherwise inaccessible places. I made an immediate distribution to my men. It was disclosed to us that Clemes and I had exhausted the stock of American spanners in Air Stores depots back home so they struggled to equip later MS and MM units adequately before they left for Australia. I cannot recall whether the trailers were fitted out for connection to mains electricity. The vans had no windows but I think the roofs were of translucent material. I don't think we were ever asked to operate at night. Our task required a small selection of ground equipment. Neither Clemes nor I had a hydraulic lifting-jack – stock shortage at home – so changing wheels would be a challenge!

I have mentioned that the airfield had been laid out in what had been virgin bush, or eucalyptus forest. The runways were to all appearances magnificent, but there was a snag. After removing the trees and undergrowth the basis of the runways was the existing gravel and sand. It was graded and rolled, topped with more sand, levelled, graded and rolled with heavy crowsfoot rollers and finally topped with no more than a couple of inches of tarred and rolled gravel. Apart from cutting back the shrubby bush beside the runways nothing was done to the ground between them, a contrast to the mown grass we always expected to see on our airfields at home. At that stage there was limited hard standing for parking aircraft. These two factors caused a real headache when the squadrons arrived in February 1945, although hard standings were increased as the year passed.

After a day or two at Nowra we became aware that another mobile unit, a MATMU or Mobile Aircraft Torpedo Maintenance Unit, was installed in its own workshops half-a-mile or so away from the airfield, beside the dirt road from Princes Highway. This unit, whose number I cannot remember, had taken over premises previously occupied by a RAAF torpedo unit early in January. It was equipped for the American 22" torpedo which was carried by the Avenger. But the strange thing was that neither the unit nor its torpedoes were ever used. We never associated with their officers who were accommodated in the Nowra wardroom. Several of these units passed through the south Pacific area but none were ever used. I took the opportunity to speak to Avenger aircrew on this subject. Without exception they told me that after going through operational training, either at Crail or in Florida, they never dropped a "fish" in anger or in practice. The torpedo range in Jervis Bay, which had been used by the RAAF Beauforts from Nowra, was never used by us. Did our carriers take torpedoes up into the Pacific? One might say that the operations in which Task Force 57/37 were engaged were not naval actions as such but land strikes or attacks on small naval craft. It was known that the US Navy High Command in Hawaii expressly forbade our carrier aircraft to attack Japanese Navy ships in Japanese home waters during the July/August operations – because the USN had a score to settle. This would not have been known to our planners when they were preparing for our participation in the Pacific war. The USN used their Avenger squadrons to attack Japanese Navy heavy warships and aircraft carriers with torpedoes, with some success.

We had almost a month to familiarise ourselves with the content of the Avenger manuals and to make sense of the parts lists. I shall comment upon our difficulties later in my story. It was during this period that I was informed that my promotion had come through so I was able to put up the second ring I had prepared for back in Edinburgh. The problem of altering my reefer jacket was solved by a visit to a tailor in Sydney. We also had to get used to the heat because it was mid-summer. Not only was it hot out on the tarmac but our vans were like ovens. Surprisingly no account was taken of the possibility of sun damage to our skins so there was no issue of slouch or sun hats and we wore our ordinary uniform caps. It was to be fifty years before I had to have both ears operated on to remove sun cancer spots. During this period we also took every opportunity to meet the people of Nowra. Due to the lack of personal transport the rules for the use of service vehicles for social purposes were relaxed and there were regular runs into the town. A commercial bus service to and from Nowra started running later, using the unsealed "back" road from the town.

I was appointed Station Sports Officer, a function which had an unusual twist. I was approached by the local Rugby League club to offer to set up a match. But in those days, and for another thirty to forty years, there was an invisible wall between Rugby Union and Rugby League. Anyone from a Union club even talking to a League club would be banned for life from playing Rugby Union. I had to refer to the Staff Sports Officer in Sydney for advice and he advised me to have nothing to do with the local club. However, I was also approached by the secretary of the RU club in Port Kembla, the industrial coastal town about fifty miles up the coast towards Sydney. He suggested that I should go up and meet him and we could spend some time together. The visit was approved and I went up on the train to Wollongong one morning, to be met by a large Scotsman with a strong Lanarkshire accent – he had been in Port Kembla since the thirties, employed by Commonwealth Rolling Mills, having migrated from Lanarkshire where he had worked in the steel industry. This rolling mill was a joint venture between John Lysaght (Australia) Pty. Ltd., an associate of the British company John Lysaght & Co., and the American company American Rolling Mills; it was built in 1936. Lysaghts had been selling galvanized corrugated steel sheet for roofing since 1857, exporting from England all around the world. They had first produced flat galvanized sheet in a factory at Tamworth, NSW, in 1880, later, in 1921, building a new mill at Newcastle, NSW. By the thirties their business in Australia was robust enough to warrant a local mill for corrugated sheet. Shared with the American Rolling Mill company Commonwealth Rolling Mill Pty. Ltd., the new joint company, was formed to build the mill on a "green field" site at Port Kembla, close to a large steelworks from where the steel billets would come. Local people continued to call the new mill "Lysaghts".

My new Scottish friend and I may or may not have talked much about Rugby football but it was his enthusiasm to show me the rolling mill that won the day. I was amazed, to say the least, never having seen a rolling-mill before. The mill was built to manufacture galvanized corrugated steel sheet for roofing, and nothing else. It must have been nearly half-a-mile long, perhaps even longer. Steel billets entered a furnace at one end at a snail's pace, proceeding down the line of rollers at ever-increasing speed, being re-heated at intervals. As the strip of steel became thinner and wider so the speed through the rollers rose until it was ripping through at some sixty m.p.h. The noise and the heat were almost overpowering. Every roller was watched by a supervisor on a gallery overlooking the mill.

Every now and again – but not necessarily on every shift – the steel would stick to a roller, causing an immediate build-up of strip. The supervisor could immediately press a button which stopped the whole mill. It was very expensive to clear the line for re-starting. Fortunately I never saw this happen. The final stages were to cut the strip, put the sheets through the corrugator and the galvanizing tanks – and they were ready for someone's roof. Before leaving the site I was shown a small museum and showroom illustrating one of Lysaght's great wartime successes, the Owen sub-machine-gun. This was an indigenous design which, after surviving the wiles of the Australian Army bureaucracy, became the weapon of choice for the Australian infantry. Lysaght made thousands of them. I had had a most enjoyable and instructive day. I don't think there was ever a rugby match with a local club.

Arrival of the British Pacific Fleet

We were aware of what the carrier task force - now carrying the title of British Pacific Fleet (BPF)- had done in the East Indies. We awaited their arrival with some excitement. Signals had obviously been exchanged about the capacities of Nowra and Schofields, another new Air Station to the north-west of Sydney, to accommodate all the aircraft the squadrons wanted to bring ashore. As a result squadrons were instructed to restrict their numbers to six each. As no time had been made available for the replacement of aircraft lost in the actions over the East Indies en route to Australia, it is possible that not every squadron flew six aircraft ashore. However, my contemporary notebook tells me that 849 *(Victorious)* and 854 *(Illustrious)* Squadrons disembarked seven and ten Avengers respectively. Perhaps some of them were what came to be known as "flyable duds", destined for immediate replacement.

We were pleased to see that, en route, all aircraft had been painted with the Pacific area blue and white roundel and bars to conform to current US practice. It had been the second change since the ships and squadrons had left the European area the previous year; they had adopted the SE Asia light and medium blue roundels while they were participating in operations in the Indian Ocean and against Indonesia.

General Motors Avenger II (TBM1) of 856 Sqn. operational training from US Naval Air Station, Squantum, Massachusetts, where the squadron formed in March 1944.

I cannot remember if it was so, but I suspect that quite a few aircraft of all types were unable to fly and were craned off the ships while alongside at Woolloomooloo. They were taken to RNAS Bankstown, an airfield in the south-western suburbs of Sydney. Great efforts had been made to start activities at Bankstown, which was to be staffed by MONAB2, comprising a Receipt and Despatch Unit and an Erection Unit. Personnel arrived in Sydney on 26th January, nearly a month after HMS *Fencer*, *Atheling* and *Striker* had brought loads of aircraft to Sydney from the UK, Ceylon and India. Complete reliance was placed on the RAAF to transport these aircraft to Bankstown and even to prepare them for flight test. The first aircraft to be tested was a Corsair II, JT537, flown by Lt (E/P) G.F. Hawkes RN on 13th January 1945. It had come from RNAS Coimbatore, India. Hawkes was the staff test pilot and he had been flown halfway around the world, via the USA and Pearl Harbor, by the USN Transport Command to get to Sydney early in January. MONAB2 started operating within a few days of arrival, making use of RAAF pilots loaned for test-flying duties.

So it was that on 10th February some sixty aircraft arrived over Nowra from the 1st Aircraft Carrier Squadron, BPF, their first sight of an airfield since they embarked from Ceylon in the last few days of 1944. This was as much of a challenge to MONAB1 Air Traffic Control as to the waiting ground crews. In addition to their normal flying crew, or perhaps even without them, most Avengers carried one or two maintenance ratings, probably not only from their own squadrons but also from the fighter squadrons. They also had to produce their own wheel chocks. It was chaos for some time! 849 and 854 were lucky to have a proper hard standing allocated to them near the control tower, but 820 *(Indefatigable)* and 857 *(Indomitable)* were parked off the runway near the intersection nearly half-a-mile away. Clemes had to cope with four squadrons, say 24 – 30, of Corsairs. To most people's surprise, two Hellcat squadrons from *Indomitable* also descended upon us.

At this stage there was no Servicing Unit for Hellcats in Australia and certainly no Hellcat parts in MONAB1 Air Stores. Men from MM1 assisted in parking this flush of aircraft. At this stage most of our MS1 time was taken up with daily servicing and the rectification of minor faults. Some experienced pilots were now allowed to leave the squadrons and return home while others were retained to help train newcomers who were brought to Australia by air. Some of the old stagers were sent on leave to

"Hellcat into the barrier". Grumman F6F-5 Hellcat II JZ935, 145/W of 1844 Sqn., HMS Indomitable, "takes the barrier" in the Pacific 1945. This aircraft was from a batch of 109 aircraft shipped direct to Australia and passed out at Bankstown

Sydney. Intensive training then began, under the control of the squadron commanders or their senior pilots. This included ADDLs (aerodrome dummy deck landings, practice carrier landings on airfield runways at Naval Air Stations) day after day. Clemes and I must have had calls for replacement wheels or tyres but nowhere have I noted the source of lifting jacks. Replacement aircraft of all three types were flown in from Bankstown by ferry pilots, some of whom were on loan from the RAAF. These "new" aircraft were almost all from Ceylon, where they had been standing in the heat for weeks and even months, unused.

Troubles began at once. The most intractable was the perishing of hydraulic system seals in Avengers. This had not been forecast as a serious problem and our stock of seals was certainly not extensive enough to cope. There was an added problem, probably not understood either on Air Stations, in the squadrons or in Air Stores. There were some quite important differences in the hydraulic systems between series of Avengers. Whether or not this was between Royal Navy Avengers Mk. I and Mk. II, or as a result of modifications, is by now almost impossible to sort out. The Mk. I was the TBF-1 made by Grumman, the designer. The Mk. II was known in the USN as the TBM-1 and it was built by General Motors at Trenton, New Jersey. In an equivalent British situation, aircraft from all factories would be the same and all the sub-contracted parts would be specified by the designing company and made or assembled on jigs and tools made by the parent company. My experience in Australia leads me to the conclusion that licence-built aircraft in the USA were not subject to such strict rules. In the case I am describing, we found that the entire hydraulic system for the Avenger retractable undercarriage could differ between one aircraft and another. The hydraulic jacks were of a different length or diameter, which meant that all pipe runs were of a different length and configuration and seals were different. As a result of exchanging old aircraft for "new", particularly in India or Ceylon, all squadrons now seemed to have a mixture of TBF-1 and TBM-1, known to all and sundry just as Avenger I or II. The replacement aircraft supplied by Bankstown could have been of either mark, but when squadrons called for an Avenger, they would receive whatever was immediately available. Whatever parts I had in my MS1 stock, I could not cover all situations – in fact, I may not have had a parts list for each mark. I discussed all this with Eric Taylor and we came to the conclusion that when indenting for a replacement jack for stock – or to repair a specific aircraft – I should quote the part number on the body of the jack and the name of the manufacturer, as well as the part number in the parts list. I continued to follow this procedure for as long as I was in charge of maintenance of Avengers. This problem had been unknown to Eric in England and he immediately saw to it that the situation was understood in the Air Stores headquarters in Sydney.

Another problem which one Avenger squadron brought from its carrier concerned the spark plugs in the Double Cyclone engines. Pilots had been complaining of rough running, particularly at full power. Fitting replacement Champion plugs from the carrier's stores made no improvement. The details now escape me, but the matter was considered so serious that a Champion representative was sent all the way from the States to sort out the problem. I think some plugs had been supplied by the UK factory of Champion which were incompatible with the original American equivalent. As soon as the squadrons had re-embarked, Clemes and I had to re-stock our spares, a procedure which gave us a real headache. Even until after the war had ended Air Stores were always struggling; I had an opportunity to visit the new building at Randwick, Sydney, which was to be the Fleet Main Air Stores, in July 1945, and it was hardly half full.

While all this hectic activity was going on at Nowra (and Schofields), officers and ratings from all the ships took staggered leave and had a whale of a time. There was also a changeover of squadron pilots so that all could have a short leave. I subsequently met an Avenger pilot who had not been required at

Nowra and, sixty years later, denied that his squadron had been disembarked to Nowra at all; he had a full fortnight's leave.

All squadrons spent a great deal of time on ADDLs and it was not long before the shortcomings of the runway construction became apparent. The small tailwheels fitted to American Naval aircraft created havoc when they broke through the thin tarred surface of the runways as the pilots dropped their aircraft the last few feet to land. Luckily the long runways enabled the touchdown point to be moved forward day by day without reducing take-off run significantly.

The squadrons from *Indomitable*, *Victorious* and *Indefatigable* departed from Nowra on 27th and 28th February to rejoin their ships, which then set off for Manus in the Admiralty Islands. *Illustrious* had been dry-docked in Sydney to rectify a vibration problem in her centre shaft so her squadrons of Avengers and Corsairs remained at Nowra until 7th March. By this time the touch down areas of runways 21 and 26 resembled ploughed fields and the airfield was to all intents and purposes closed while remedial measures started. This affected the immediate future of 723 Squadron and the tidy dates quoted in certain books are questionable. 723 was supposed to have formed at Nowra at the end of January for Fleet Requirements Unit (FRU) purposes. I don't think that was the case as I cannot remember the Martinets being there when the carrier squadrons arrived. It is known to have spent some weeks at Bankstown, where its Martinets had passed through the Receipts and Despatch Unit, or the Erection Unit. As mentioned above 723 had had temporary charge of the flight of Expediters which had started to operate from Mascot, Sydney, by the end of 1944. It is my belief that the Martinets did not take up residence at Nowra until the runways had been repaired. There was no urgency for them to be based there while there were no first-line squadrons ashore and 706 Squadron had not yet been formed. 723 operated from Jervis Bay for a short time during May and finally settled into Nowra towards the end of that month. It is notable that when *Formidable* came to join the Fleet and went alongside in Sydney between the 1st and 22nd June, 848 Squadron's Avengers and the Corsairs of 1841 and 1842 Squadrons disembarked to Jervis Bay and not Nowra. MONAB5, *HMS Nabswick*, is said to have commissioned at Jervis Bay on 28th April 1945 but the Station was by no means fully built by then – and was never completed by the time the Royal Navy withdrew in October. I suspect that both MONAB1 and the Fleet staff in Sydney were so concerned about Nowra's runways that they needed as much time as possible for adequate repairs to be completed before squadrons could re-start ADDLs. Although in theory all facilities should have been available at Jervis Bay, I surmise that while Jervis Bay airfield was used, all officers and the few ratings who came ashore there were accommodated at Nowra where there would have been plenty of space.

Many of us at the Air Station had taken the opportunity to meet and get to know the people of Nowra and many friendships were formed. We learnt how a few of the local boys were serving with the RAAF overseas, some in the squadrons based in England. Their parents were very hospitable and, of course, were very interested to hear about our home country. We experienced for the first time the typical Australian bungalow, seen all over country areas of the continent. It was lightly built around a timber frame, usually on a plinth which could be as low as one foot or as high as four or five feet, in the latter case providing stowage for wood for the hot water geyser. There was usually a verandah and the whole structure was roofed with galvanized corrugated steel from Port Kembla; this construction recognised the benign climate of this part of the continent. (It is possible that the outer cladding of the building was also of corrugated sheeting.)

Avenger I or II 7P of 857 Sqn taking off from HMS Indomitable somewhere in the Indian Ocean before sailing for Australia. Note Indian Theatre markings

With the departure of the squadrons from Nowra neither Clemes nor I had any useful role to play. However, within a few days we learnt of the formation of 706 Squadron at Jervis Bay with a nucleus of two Avengers and two Corsairs. The squadron was officially forming at Schofields, west of Sydney. Its ratings had been trooped out, arriving in Sydney early in March. The story is tortuous, but I shall try to explain. Clemes and I were ordered to take our MS units down to Jervis Bay on 1st or 2nd March to do the maintenance for the new squadron. There were no huts available so our MONAB sent down a two-man tent for Clemes and me; a Matador was summoned to tow our four trailer vans to Jervis Bay. A Nowra party also brought down and erected the MONAB portable hangar. Our lorries provided workshop bench facilities. A place was found for the tent in a small area of green sward and it was furnished with two beds, two chests and two chairs. A few yards away from our tent was what seemed to be the only hut on the airfield which had electric light, running water and civilised plumbing. There was no accommodation for our men who continued to live at Nowra and commute in one of the Bedford 3-tonners. The hut was designated for officers' messing and had a bar. I suspect that our first pilots found beds in the Jervis Bay "village". A major surprise for us in our first evening there was to find that there were already three lieutenant commanders in residence; L/C (A) R.E. Bradshaw DSC** RN, L/C (A) D.M.R. Wynne-Roberts RN and L/C (A) C.A. Fraser RN. All were accommodated in the village. Now we were there, what could we make of Jervis Bay, or "JB" as it came to be known? (See earlier map for location of Jervis Bay.)

One should not consider JB in the terms of RN Air Stations as we knew them in the United Kingdom at the time. JB was beyond the back of beyond, about thirteen miles south-east of RNAS Nowra, along a dirt road off the Prince's Highway through unbroken bush of 20 – 30ft. trees, about three miles from the ocean and a mile or so from Jervis Bay. The bay was to the north of the airfield. This two-runway airfield had literally been hacked out of the bush. We were to understand what that meant when a bush fire started up between the runways; we were only saved by the runway intersection where it burned itself out. The fierce crackling noise and the sight of whole trees going up in flames across a hundred-yard strip were extremely unsettling. Both runways were approximately 1,500 yards long and 50 yards wide and were of the same construction as Nowra. The public road led to both the airfield and to the JB "village".

On the inland side the village was fenced and the entrance gate had a sign on it "A.C.T." We learnt that this stood for Australian Commonwealth Territory - but what for? This was the original home of the RAN (cadet) College, *HMAS Creswell*, founded in 1913 and built on virgin land. It comprised a number of buildings, many facing a cricket oval. The Depression hit Australia hard and in 1930 the site was evacuated and the College re-established in the Flinders Naval Depot, Melbourne. There was no civilian activity in Jervis Bay to occupy the buildings so the whole establishment was let out in parts for use as a holiday resort. For years it attracted young people from Sydney as a get-away holiday of sun and sand – and other things. There were no leisure activities such as one would expect nowadays in a seaside resort. The site was returned to Naval use in 1958 when the College returned from Melbourne. It has since been expanded and rebuilt and is now an impressive establishment. If the holiday resort had not been there in 1945 it would have been almost impossible to operate the airfield. Foley's Hotel was requisitioned for senior officers and half of Links House accommodated junior officers. A pleasant hut in the bush on the edge of the establishment known as "The Cottage" was used as the airfield wardroom. The whole place had been well cared for by the occupants, including the planting of flowering and ornamental shrubs. It was a haven for wildlife. The "laugh" of the kookaburra bird could be heard frequently and there was the flash of colour of the crimson rosella parrot to delight the eye. It was not uncommon to encounter a wallaby on the road through the bush from the Prince's Highway. JB, the airfield, the College and the adjacent land occupy a peninsula three parts surrounded by the sea. It must have its own micro-climate because an area on the south-facing slope down to Sussex Inlet is now fenced off and designated the Booderee National Park and Botanic Garden. The bush has been opened out and new planting of indigenous trees and shrubs has been carried out. Further out, on the southeast side of the airfield, a large area has been declared Jervis Bay National Park. To the south and along the coast new resorts have been developed so JB is no longer at the back of beyond, as we found it in 1945.

As I write this in January 2021, a huge area of bush in the vicinity of Nowra and Jervis Bay is burning furiously and many homes in the vicinity of the Bay, Sussex Inlet, where there had been a small RAAF seaplane station in 1945, and in the area of St. George's Basin, have been reduced to ashes. RANAS Nowra has been used as a helicopter base during the emergency.

I do not know the exact details of MONAB5's occupation of JB airfield except to say that when their personnel first came down from Sydney they were put up at Nowra for nearly two months from the beginning of March to the end of April, indicating how the development of JB as a proper base was struggling. There was no other regular Naval flying during the short period we were there with 706. To illustrate how basic it all was, the day following our arrival there was a heavy downpour in the afternoon. After a very short time we suddenly realised that the sward on which we had pitched our tent was a major point of drainage off the airfield. Within minutes the water was flowing through the tent and immediate evacuation was necessary. We had to manhandle our furniture through the rain and flood to the mess hut where we spent the night. The following day we moved into one of the JB hotels. Heavy rain several days later accompanied the forced landing of an old De Havilland Dragon biplane which was flying a regular passenger service up and down the coast from Mascot. One of the Gipsy Major engines was misfiring so, with few other landing grounds on the route the pilot made a precautionary landing at JB. One of our AE fitters had a look at the ignition system and found one of the spark plugs to be dead. The plugs were of 18-mm thread; both the Pratt & Whitney Double Wasp and the Wright Double Cyclone fitted 18-mm plugs so they were all we had in our spares. Miraculously, they worked, so we fitted a set of eight (two per cylinder) and away he went – cost free!

Also amongst the memories of JB is the brief story of a misbehaving Double Wasp in one of our Corsairs. Clemes' boys couldn't find what was causing the engine to run roughly and not develop full power. We summoned one of the team of Staff AEOs from Sydney to help us out. He asked us to prepare for a full-throttle ground run. The tail was tied down to concrete blocks and off he went. A big engine at full power is an awesome sight and sound. The aircraft gave the impression of being alive as it bounced and rocked at nearly 2,500-rpm Suddenly, after no more than half a minute there was an impressive explosion low on the port side of the engine, accompanied by an ominous bulge in the engine cowling and a sheet of flame. Appreciating the situation, the AEO had the courage to stay in the cockpit until he had switched the engine off and turned off the fuel. He then made the quickest exit from the cockpit I had ever seen and ran for safety! Our aircraft was fortunately still in one piece. We were all rather chastened as we waited for things to cool down. Subsequent examination showed that one of the lower rear cylinders on the port side of the engine had parted company from the crankcase, leaving its attachment flange in place but breaking all the induction, ignition and exhaust connections and the drives to the valve gear. It would be interesting to know what happened to the aircraft after that.

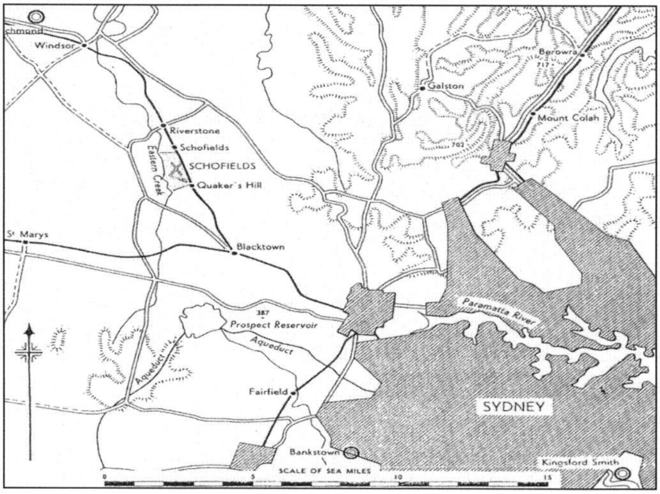

Map showing position of Schofields and Bankstown in relation to Sydney.

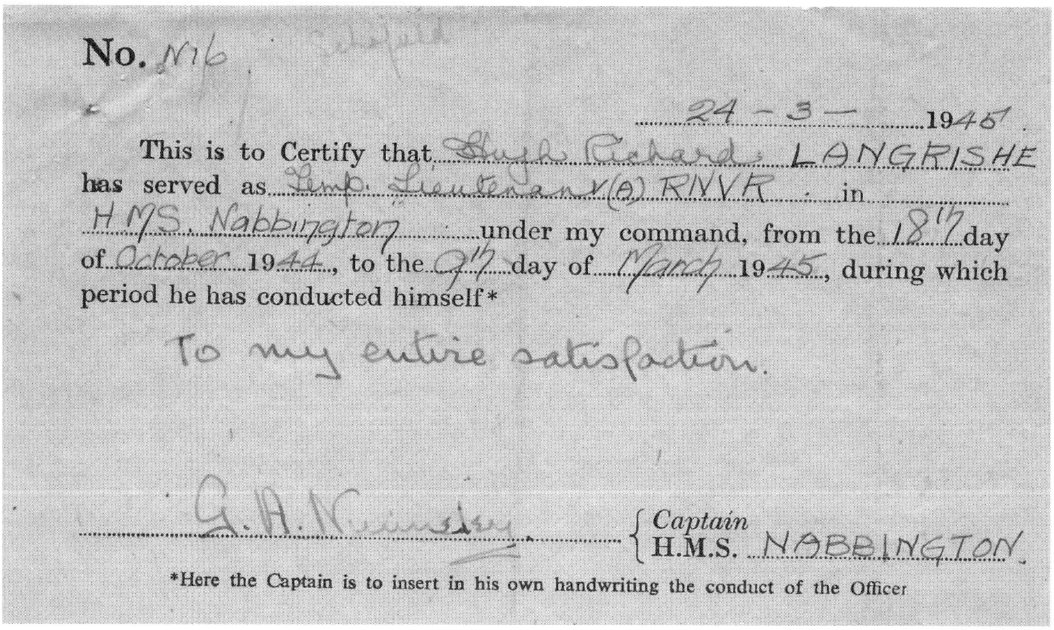

No. N16

24 - 3 - 1945

This is to Certify that Hugh Richard LANGRISHE has served as Temp. Lieutenant (A) RNVR in H.M.S. Nabbington under my command, from the 18th day of October 1944, to the 9th day of March 1945, during which period he has conducted himself*

To my entire satisfaction.

G. A. Nunneley — Captain H.M.S. NABBINGTON

*Here the Captain is to insert in his own handwriting the conduct of the Officer

Flimsy from Cdr. Nunneley at Nowra on my departure for Schofields, indicating the date when I was removed from their Station establishment - 24th March 1945

RNAS Schofields

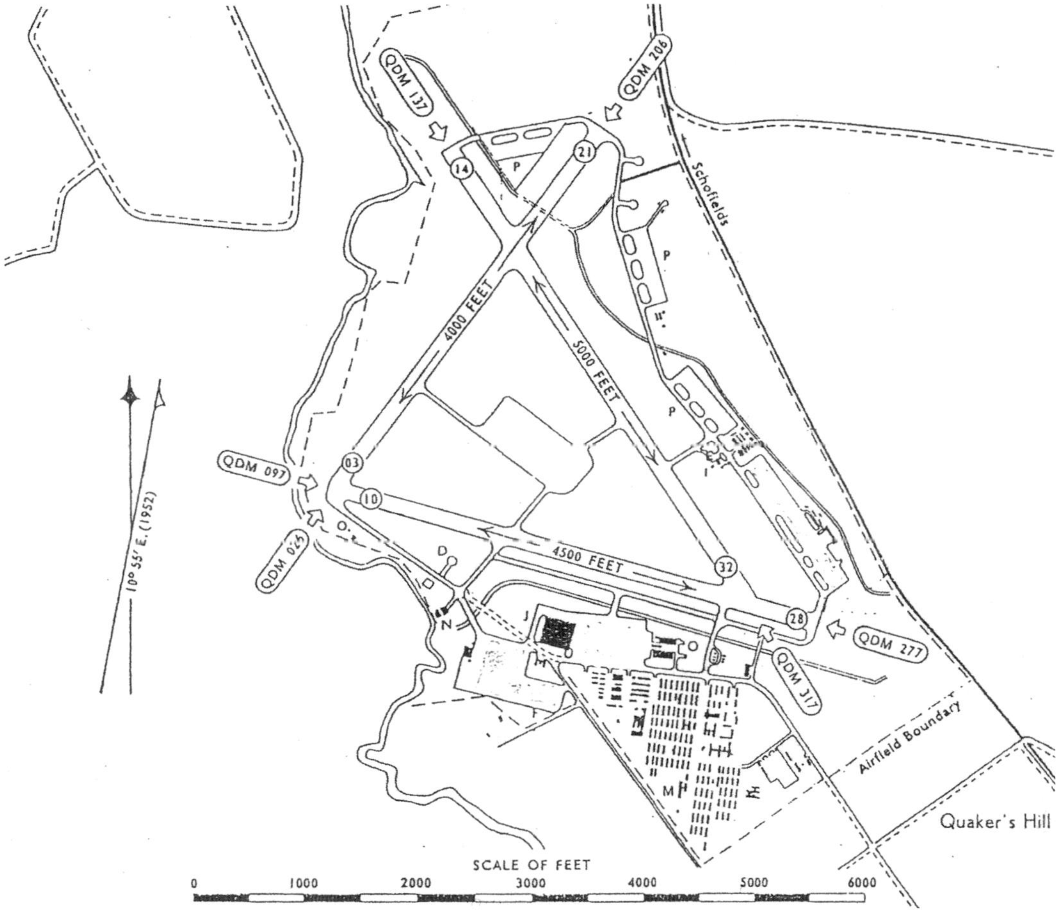

RNAS Schofields was taken over in January 1945. The technical area only had one hangar. All squadron aircraft were parked in the open on hard standing areas. This plan is dated post-war and shows many hangars which did not exist in 1945-'46

Within a week of our arrival at Jervis Bay the three lieutenant commanders disappeared in the direction of Sydney, Bradshaw to Schofields to take command of 706 Squadron. The three serviceable aircraft were flown off to Schofields, whereupon by verbal order Clemes and I prepared ourselves to follow with MS1 and MS2. This was not an easy move as there was only one Matador available to tow our four trailers, each tow occupying a day for the return journey of two hundred miles. We both used our Bedford trucks to transport our men and personal belongings. We had no maps and none of us knew exactly where Schofields was. But local Australians knew all about us and we did the drive without problems – in fact it was a very interesting route, first up the coast through Wollongong where we stopped for a prawn lunch, then up the Bulli Pass and over the Illawarra Coast Range. This is the area from which Bulli soil is taken for surfacing all the important cricket squares in NSW. We must then have aimed for Parramatta then followed the railway via Blacktown to Schofields. The journey was slow and hot. We were now towards the end of the southern summer and this showed in the dry countryside. There is absolutely no resemblance to the countryside of the lowlands of the British Isles. In Australia the gum tree leaves are the same colour all the year round, but grassland and scrub have turned to a uniform dun colour. This was particularly marked in the vicinity of Schofields. Our aerodrome was between Schofields settlement and Quaker's Hill. It was about twenty-five miles north-west of Sydney centre. There was a direct rail route into the city, with Quaker's Hill station only some quarter-of-a-mile from the main gate of the Air Station. This was a branch line to the country town of Richmond, to and from which there was a regular passenger service with corridor coaches drawn by a steam engine. The NSW State Railway issued all Naval personnel with fortnightly free passes for travel throughout the suburban railway network, of which this branch was a part. It was a well-used concession.

Schofields was another airfield planned for the RAAF. The land was open scrub and bush. The runways were laid in a triangular pattern and were 1665, 1500 and 1335 yards long and 50 yards wide. Technical information supplied to the Royal Navy was dated May 1944 and was incomplete. The runway construction was specified as of unsealed gravel, but they had been sealed in the same way as Nowra by the time the aircraft arrived from the Fleet in February 1945. Pondering over this business of runway construction, it now seems to me that in all three cases mentioned, there were just no sources of asphalt or concrete in sufficient quantity within a reasonable distance to enable runways to be completed to a higher standard. The carrier aircraft which arrived in February were the Fireflies and Seafires from *Indefatigable*. They found the airfield facilities and the camp far from complete. Indeed, the wardroom hut had not even been started. By that time MONAB3, *HMS Nabthorpe*, had been in occupation since mid- to late January and technical ratings were employed to help the contractors assemble hangar doors, probably amongst other things. As was the case at both Nowra and JB there were no hangars on the squadron dispersals. This made for the acute discomfort of the maintenance ratings as the summer sun rendered the aircraft surfaces untouchable. The technical Mobile Units of MONAB3 were equipped and stored for Fireflies and Seafires and helped to look after the squadron aircraft while they were ashore.

A week to ten days after the embarkation of the squadrons and our arrival at Schofields, the Station was still noticeably incomplete. The wardroom huts were all fitted out and in use but the officers' ablutions and heads were still unfinished. While the equipment was all in place there was no privacy; WC pans lined both sides of their hut, open to view to all who entered. Happily, it was not long before partitions were installed and one could commune with nature in privacy. It was to be several days before the four trailer vans of MS1 and MS2 were delivered from JB and parked in the technical area, a couple of hundred yards from the 706 dispersal. We took it as obvious that we should continue to work with 706 Sqn. because it was then the only unit there and therefore the only one with Avengers and Corsairs. The dispersal they occupied was on the north-east side of the airfield at the south-eastern end of runway 32. The role of the squadron was quite clear. It was to provide aircraft to keep spare pilots in flying practice until they were sent up to the carriers and, when necessary, to train pilots on other types. Training in all aspects of the front-line use of both fighters and bombers was to be an important aspect of the squadron's work. But during most of those hectic months before the August ceasefire the frequent movement of pilots and observers in and out prevented the implementation of a comprehensive programme. Under Lt Cdr Bradshaw the squadron now had two senior pilots, Lt (A) Peter Jones RNVR was the fighter instructor and Lt H.G. Thom RANVR was in charge of the torpedo-bombers. The Senior Observer was Lt (A) Alan Clayton RNVR. One could not have asked for a more congenial team.

The squadron's aircraft gradually increased in types and numbers, acquiring Hellcats, Seafires and Fireflies at first, then, in August, when the Light Fleet Carriers joined the Fleet, the Barracuda II was added. During that period numbers of aircraft were gradually increased to six of each type. Also during the quick-moving events of these few months Stations were issued with aircraft for a Station Flight. Initially this meant that each flight was issued with two Tiger Moths, transferred from the RAAF and carrying RAAF serial numbers. I cannot reconcile the reference books with what actually happened regarding command of the squadron. I am reasonably sure that Bobby Bradshaw was superseded by Wynne-Roberts as early as July 1945, and Angus Fraser, who I believe had been Staff Officer with 899 Sqn. when they were training RAAF pilots for carrier operations at Schofields, took over before 706 transferred to Maryborough at the end of August. During this time we were still struggling with aircraft which had come out of storage in India and Ceylon while Bankstown and Archerfield were able to issue new aircraft to the first-line squadrons. It was at Schofields that I really got to know the "Chiefy" of the squadron's maintenance ratings. He must have been one of the last of the RAF Flight Sergeants still serving in the Fleet Air Arm. He had been a career technician since the 1930s and had been with the FAA throughout the war. He told me he preferred it that way! His previous posting had been as Chief in 1820 Sqn. which had been the unfortunate guinea pig issued with the Curtiss Helldiver (the version built by the Canadian Car & Foundry Co. and known as the SBW-1.) The story of this episode is well told in the book "Curtiss SB2C HELLDIVER" by Peter C. Smith (The Crowood Press, 1998). "Flight" was a delightful man and ran 706 Sqn. very efficiently, he didn't really need an AEO! I wish I could remember his name. The squadron was the only occupant of its dispersal area.

I have already mentioned that there was no hangar in which to carry out such routine things as Minor Inspections. I cannot recall ever doing an engine change. Several huts sufficed for the administration of the squadron, one of which must have been where the ratings kept their precious toolboxes, and from which "Flight" organised his men. The CO and his staff shared their hut with the aircrew crew room. One can judge the size of the squadron's dispersal by the fact that when at its maximum establishment of thirty-six aircraft they could all be parked together – and Corsairs and Avengers were usually parked with their wings folded.

Supermarine Seafire L.III NF452 on completion by Westlands at Yeovil, Jan. 1944. This aircraft passed through the hands of 885 and 807 Sqns., latterly embarked on HMS Hunter. seeing action with the East Indies Fleet in 1945. In the background is one of the first production batch of the Westland Welkin 1 high altitude fighters, in the DX series, possibly DX325. Although built in some quantity none were issued to front-line squadrons.

For the six months until the end of August I remained in charge of MS1, independent of any other unit at Schofields. My eight or nine men were still my divisional responsibility although for most of the time they were attached to 706. This was the obvious thing to do because, apart from a four week period in June when the Avengers of 820 Sqn. disembarked from *Indefatigable* with the ship's Seafires and Fireflies, i.e. the whole Air Group, the only Avengers at Schofields were in 706. This was a time before the concept of a Carrier Air Group (CAG), first adopted many years before by the USN for all its carriers, had been fully adopted by the Royal Navy. Wherever it went, a USN CAG kept together, ashore as well as afloat. Unfortunately, the organisation of the MONABs was based on the idea that there would be bases which specialised on individual types of aircraft – thus the MS units would identify their original Bases as only suitable for specific types, splitting the Air Groups as they flew ashore. To be fair, it was recognised that the MS units, and probably the MM units, could be moved around, as happened to MS1 and MS2. But if movement was to be carried out on a large scale, the problems for Air Stores would be acute. At Schofields I found my work with MS1 to be very largely as a storekeeper. How I did it I do not remember, but I found my way into the Air Stores administration offices in the middle of Sydney, by-passing the Air Stores department at Schofields, and I even obtained spares quickly by doing so. Now we had settled down at Schofields and integrated our work and spares into 706 Sqn., I found the demand for parts widened, particularly for small items such as nuts and bolts and all manner of what we have known as Aircraft General Spares (or AGS) parts. All British industry used the standard Air Publication for AGS parts and all aircraft type parts lists referred to it. The American system allowed the manufacturers to introduce their own lists of standard parts, and the USAAF and USN did likewise. The numbering was all different even though there was interchangeability. My contact in Sydney had been involved in Air Stores for the Americans when they were in Australia before us, and he knew the local subsidiaries of the US engineering companies such as General Motors. GM made the Holden car there and had a number of sub-contractors who concentrated on the small bits and pieces of general spares. During the war they turned to manufacturing parts for the aircraft developed and built by The Commonwealth Aircraft Company based on the NA Harvard. Through this labyrinth I actually got what I wanted.

There is no doubt that we encountered many problems while keeping an adequate number of aircraft serviceable for a meaningful flying programme. None of those issued were straight from the factory, and I can confidently say that most of the replacement aircraft issued to the front-line squadrons in February and March were re-circulated from Storage. 706 Squadron's Avengers were all Mks. I and II at this stage and all the squadron's aircraft were finished in the standard three-colour sea camouflage of the FAA. I mention this because of what was to come later in the year. In wartime terms all these aircraft were "old", probably up to two years in age but under 100 flying hours. Modern air engineers might be horrified to hear this, but it was a fact that very few aircraft reached their 240-hour Major Inspection, let alone the time for overhaul at factory or Naval Repair Yard. It was now that 706 Sqn. experienced the irritating business of the incompatibility of hydraulic system parts already mentioned.

I had to cope with an excitement one day when called to the crash of a 706 Sqn. Avenger only a few hundred yards north of the airfield boundary. The pilot had done an excellent wheels-up landing amongst open scrub when the engine failed after starting to pump out a dense cloud of black smoke from both exhaust pipes. I found that there was no damage to the upper side of the aircraft and all the top engine cowling panels were accessible. I reported my observations to MONAB3's Cdr A/E and, without delay, and understanding the consequences of a repetition of the failure, he asked me if I would carry out an investigation where the aircraft then sat. There was no difficulty in getting light transport on to the spot so, accompanied by two of my engine fitters and plenty of tools, we made a start. Removing the cowlings over the rear of the engine we could see that there was oil everywhere. It was obvious that large quantities of engine oil had suddenly been released from the rear of the engine, some entering the induction system through the supercharger. This led us to start at the beginning – the engine-driven centrifugal supercharger situated between the carburettor and the induction tracts. Luckily for us, this was a down-draught system so by removing the carburettor we had immediate access to the induction side of the supercharger, without getting ourselves covered in oil. We removed the magnetos and other accessories driven by an extension shaft through the supercharger rear bearing. This disclosed the supercharger rear cover, a light alloy casting, and we could then see that the casting had disintegrated around the bearing housing, releasing engine oil at full pressure; there was then nothing to stop the oil from being drawn into the engine by the supercharger, momentarily being burnt in the cylinders with the fuel until the sparking plugs were fouled and the engine stopped.

North American SNJ-4 Texan, US Navy basic 2 set trainer version of the Harvard, used by the Royal Navy flight cadets in the US Navy training schools in Florida

Not having seized, the engine continued to rotate, pumping out clouds of oily smoke until the aircraft finally struck the ground and brought the propeller to a halt. This investigation took no more than a morning to complete and we could release the aircraft for the salvage party from Bankstown to collect. I still had a report to write but I was satisfied that we had found the answer to the problem. This analysis was included in the subsequent Form A25, the official accident report.

706 Sqn. gradually increased in size during April and May, keeping pace with the replacement aircrew who came from the UK, Ceylon and the USA, the latter stream from the US Navy training schools in Florida. At some time in 1944 pilot courses arriving in Canada from the UK had been split in two, half remaining in Canada, the other half going to Pensacola, Florida. Basic and advanced training were carried out on one or other of the numerous satellite fields of Pensacola, on Stearman N2S biplanes for basic training, then North American SNJ Harvards. These men continued to operational

type training within the Jacksonville group of airfields on the Florida peninsula, having been separated into fighter or torpedo-bomber pilots after passing out at Pensacola. This would seem to have been a satisfactory system because there were by that time many squadrons of Avengers, Corsairs and Hellcats in the FAA. Avenger pilots were introduced to torpedo dropping, using the American Mk. 13 22in torpedo – something they were never to do again. On return to the UK or, latterly, to Australia, integration into British routines and organisation took place in squadrons such as 706 which could undertake armament and operational training. There was no time for posting to other second-line squadrons to gain flying experience. This had been common practice for pilots returning to the UK from Canada, although not universal. USN- trained pilots received their wings in Florida, were commissioned and obtained their uniforms there. We were very impressed by the quality of their uniform suits and tropical khaki. Training in the southern states had one drawback: it took longer than other training programmes. Pilots following the Canadian stream were in front-line squadrons as much as four months before men who had started out in the same Naval pilot's course in the UK but had gone to Florida.

There were two other long-term resident squadrons at Schofields. 1772 Sqn. arrived with their twelve Firefly I in March after transit in *Ruler* from the UK. They continued working-up until they replaced 1770 Sqn. in *Indefatigable* on 7th July for the operations off Japan through to August and the end of the war. Also on board *Ruler* was 885 Sqn.'s 24 Hellcat I and II. They stayed with the ship which was to provide air cover for the Logistic Support Group up in the Pacific during Operation *Iceberg*. This was part of the US Navy operation to capture Okinawa. The BPF First Aircraft Carrier Squadron operated as Task Force (TF) 57 to suppress Japanese aircraft activity from Formosa and the island chain of the Sakishima Gunto during April and May 1945. *Ruler* also provided Combat Air Patrol (CAP) and a spare deck for the Task Group supporting the operation against Truk by *Implacable* in June. The Hellcats were joined by several Avengers in May for anti-submarine patrols and by Corsairs in June. By this time the squadron was providing replacement pilots and aircraft for the Fleet Carriers as well as carrying out its air protection duties. The other long-term resident squadron at Schofields was 899 Sqn. They brought twenty-four Seafire LIII with them from the UK aboard *Chaser* which arrived in Australia on 23rd April. The COs of the Seafire carrier squadrons had been concerned at the poor quality of replacement pilots so it had been decided to fill the gaps with experienced RAAF Spitfire pilots. The first course of some twenty pilots (the number has been officially put at twelve, but my memory tells me that there were more than that), from squadron-leader rank downwards, many with the ribbon of the DFC (Distinguished Flying Cross). They were to be inducted into Naval practices by 899 Sqn. which was re-classified as an OTU, under the same command of Lt Cdr (A) G. ("Shorty") Dennison RNVR. Several experienced pilots were posted in to help as instructors, including S/Lt (A) R.H. ("Dickie") Reynolds RNVR of 887 Sqn. in *Indefatigable*. He had shot down three Japanese Zero fighters during Iceberg and was considered an exceptional pilot. Some intensive flying then took place, including a large amount of dummy deck landings which partly aimed to change the habits of hours of gentle landings on lengthy runways by the Air Force pilots. Later in the course a demonstration was given to senior officers of the Royal Australian Navy, into whose Volunteer Reserve the RAAF pilots were transferred, all as Lieutenants. Unfortunately, one of these young men lost his life in a crash from low level during training.

Fairey Firefly FR I MB590. This aircraft is a Mk. NF I of 1791 Sqn., a night-fighter squadron, but the type was issued to 1772 Sqn. in 1945, also to 706 Sqn., as FR I. The ASH radar sets became widely used at the end of the war and not just for night fighters

I have mentioned how since our arrival in Australia we had been struggling with old aircraft drawn from storage in the UK, Ceylon or India and brought to Australia on Escort Carriers or Maintenance Carriers. New Corsair squadrons in the theatre, 1845 in February ex *Slinger* and 1843 ex *Arbiter* at the beginning of May, had both been re-equipped with twenty-four Corsair IVs (FG-1D) before they left the UK on passage. These new aircraft, from a production line set up by the Goodyear Corporation to build equivalent versions of the Chance Vought F4U, had been delivered finished in the current USN all-over Midnight Blue paint, (some reference books quote this colour as No 623 Glossy Sea Blue) replacing the standard RN camouflage in which all previous marks of Corsairs had been delivered. 1845 had been split on its arrival at Schofields to replenish the Corsair squadrons on *Formidable* and *Victorious* with both men and machines. 1843 Sqn. stayed ashore as a spare squadron, was moved to Maryborough and finally Nowra, lost its aircraft there and was shipped home to be disbanded in the UK on 10th December 1945. We had also seen some blue Hellcat II from 885 Sqn. *(Ruler)* in March. In spite of this, we were still being issued with re-furbished early marks by Bankstown, ex-storage, although Archerfield (TAMY1) had opened a preparation line of Corsair IV in April/May, in addition to Seafire III and XV. Unfortunately, we were never in the position to issue the latest new aircraft to all first-line squadrons in the BPF during the war, as was the case with the USN. However, we were doing our best to hasten the process.

It is perhaps appropriate to illustrate the difficulties of planning squadrons for the BPF in the United Kingdom by describing the fate of 1840 Sqn.'s Hellcat II and their pilots. The squadron had been allocated to *Speaker* in December 1944 as one of the four constituent squadrons of the new 3rd Naval Fighter Wing, each embarked on its own carrier. They sailed separately for Ceylon where two ships were retained for the East Indies Fleet. *Ruler* followed, as described above, to Sydney, a month behind *Speaker*. The latter was the first to join the Fleet Train at Manus and sail with the train for *Iceberg1*. Immediately a number of pilots from 1840 Sqn. were detached and joined 1839 Sqn., Hellcat II, on *Indomitable*. 1840 Sqn. continued to work with *Ruler* on CAP during refuelling but when the Fleet Train withdrew to Leyte Gulf after the conclusion of *Iceberg1* the remaining pilots of 1840 were transferred to 1839. 1840 squadron was formally disbanded on 27th April 1945, leaving its maintenance ratings in *Speaker* to look after the spare aircraft the ship was carrying as replacements.

Nowhere have I seen a discussion of the theory and practice of protective colouring, or camouflage, of aircraft habitually flying over the sea. At no time during the war until 1944 was there any similarity between the USN and RN colour schemes. Although both Services seemed to agree that the undersides of aircraft should be different from the upper sides, there was no common colour and there was certainly no agreement over the upper sides. Until early in 1944, when the USN abandoned all vari-coloured schemes, aircraft for the Royal Navy were delivered in British camouflage, even if the aircraft were released from USN allocation under Lend Lease. From early 1944 all USN carrier aircraft were finished in an overall Midnight Blue. A decision must have been made by the British Purchasing Commission, with approval from London, that we would accept all Naval aircraft in the new standard USN scheme, but with British roundels. This decision coincided with the release of the first FG-1D (Corsair IV) to the Royal Navy, so all blue Corsairs were Mk. IV. In the Pacific theatre the roundel was similar to that adopted by the US air forces, i.e. a large blue roundel with either a large white star centre (US) or with a narrow white outline and a white centre (RN), both having white bars at each side. The tri-colour flash on the fins of our Naval aircraft was removed. A decision to ship American aircraft to Australia must have been made in mid-1944. British Air Commission figures show that the USN transport carriers moved 80 TBM-3 Avenger III, 207 FG-1D Corsair IV and 122 F6F-5 Hellcat II to Australia during 1945.

Grumman F6F-5 Hellcat II JX772 etc. lined up at Grumman works for test. This batch were shipped to the UK. All new Hellcats at this time were finished in midnight blue and finished with northern theatre markings. JX772 was in action in the Pacific with 1844 Sqn.

The Corsairs would be those photographed in preparation in the large hangar at Archerfield in 1945. Being of the latest marks, all these aircraft were finished in midnight blue. It is of course possible that some of the later ferried aircraft were never landed, as Lend Lease terms required us to return or destroy unwanted aircraft as soon as VJ-day arrived. From this explanation one can see that it was possible to date American aircraft by their colour scheme. I don't think there was one blue Corsair amongst the four squadrons which came ashore to Nowra in February; they were still operating Corsair I and were augmented by Corsair II in February/March 1945. The replacement aircraft which might have been needed from time to time while the fleet was based in Colombo would have come from Coimbatore and would have been stocks delivered direct from the US. Some of these were brought on to Australia. There was a handful of blue Hellcat II from *Indomitable*; replacement aircraft seem to have been mixed because both squadrons retained some Hellcat I until they disbanded later in the year.

Interestingly, some Hellcat II were delivered before the change in colour, but broadly speaking a blue Hellcat was a Mark II and a camouflaged one a Mark I. Delivery of aircraft from the States before the war ended took two forms. Most multi-engined aircraft for the USAAF and flying-boats ("seaplanes") for the USN, were delivered by air, staging through the Pacific islands in convoy. Some attempts were made to do the same for single-engined aircraft, but the risk of loss was too great. With the astonishing speed of building of escort carriers sufficient became available to set up a special USN transport squadron in April 1944 – Carrier Transport Squadron Pacific – comprising thirteen escort carriers and two aircraft ferries. By mid-1945 there were twenty-five escort carriers, plus the temporary addition of six RN escort carriers between December 1944 and August 1945. Were the latter used to transport the aircraft for the Royal Navy whose numbers are mentioned above?

Off Duty in Sydney

Life on the camp at Schofields was still pretty basic. There was undoubtedly some building work still going on but photos taken in August when the Light Fleet carrier *Glory* disembarked her squadrons show a small collection of tents beside the aircraft dispersal. The lack of green countryside outside the airfield stamped its character on the whole area; Nowra was a more attractive place with the eucalyptus forest not far away. Morale remained high at Schofields very largely because of the proximity of the railway into Sydney. There were only two stops, Blacktown and Parramatta, so the train could maintain a fair speed into the City centre, terminating at Sydney Central. It was quite practicable to go into Sydney for the evening – but beware the licensing rules. The only place to buy a beer after 6.00 p.m. was a restaurant or club, encouraging the "six o'clock swill".

I was able to get to know several more of the families whose names and addresses I had been given before my departure from home. One was the delightful family of a retired Australian general, Gen. Miles, who lived at Vaucluse, one of the Sydney suburbs on the eastern side of the harbour. Their house was on sloping ground looking westwards across the Harbour to the bridge in the distance. It became almost a home-from-home for me and I spent many happy weekends there, often staying the night. To help entertain me there were a son and daughter a year or two younger than me. We played tennis at the local club, swam in the sea and took tea and evening drinks in the garden. It was unashamedly typical middle-class English living! And to get there and back from the city I travelled on the famous Sydney "toast rack" trams. Another of my hosts took me to Sunday lunch at the Royal Sydney Golf Club. This was really a Sunday parade; all Royal Navy officer guests were expected to be in uniform. I saw a familiar face with four rings who immediately addressed me by my name. He was Capt. McCarthy RN and he had been the Commanding Officer at St. Merryn two years before when I had been doing my qualifying course. We had only spoken twice then. I was most impressed. Also, it was an excellent lunch in beautiful surroundings.

During this period we also experienced VE-day. With all our eyes focused on the goings-on of the BPF, it somehow did not have the impact of VJ-day three months later. We did break from our daily routine, however. Shortly after breakfast on that day we were alerted by the noise of aircraft and hurried out to see a formation of a squadron of twelve RAAF CAC Boomerang fighters flying over the airfield. I believe they made for Sydney and flew under the bridge – but this may just be hearsay. As the trains were still running, many of us set off for Sydney and gathered in Martin Place to celebrate. This is where there is a cenotaph commemorating the Great War and it is usually the focal point for such gatherings. After it was all over and a quantity of beer had been sunk we returned to Schofields to confront reality. By now the southern winter was approaching and we changed into warmer clothing. I showed my continuing loyalty to the MONAB by wearing my khaki battledress during the day.

It was also the beginning of the rugger season. I had brought my boots and a strip and kept an eye open for opportunities for a game. There must have been a sports officer in our HQ and some liaison with the RAN because, replying to a notice in the wardroom, a small number of us were invited to go to the football ground on the King's School, Parramatta oval, for a "pick-up" game. King's is a rugger school and I remember that during the winter of 1938 the school had sent a touring team to England – that would have been partly in their long summer holiday, and they did very well. We had no football of either code at Schofields, so we had no inter-Station competition. But the RAN were involved in their own and a series of tri-service matches.

At North Sydney Oval

1.15 p.m. Referee: Mr. J. A. O'CONNOR.

Royal Navy v. Royal Australian Navy

ROYAL NAVY (White).

Full-back: (1) A.B. Boyd.

Three-quarters: (2) Lt. Langrishe, (3) F. Sgt. Reynolds, (4) C./Yeo. McCall, (5) S./Lt. Burroughs.

Five-eighth: (6) C.P.O. Bettinson.

Half-back: (7) S./Sgt. Lowe.

Forwards: (8) L./Cook. Casey, (9) L.A.F. Berry, (10) Lt. Henderson, (11) P./O. Bickford, (12) Lt. Tarrant (Capt.), (13) S./L. Boyes, (14) M.N.E. Lewis, (15) A.B. Woodburn.

Reserves: (16) S./Lt. Norman, (17) S./Lt. Syson, (18) L.R.M. Coombes, (19) L./Te[l.] Foley.

ROYAL AUSTRALIAN NAVY (Navy Blue).

Full-back: (1) A.B. Falla.

Three-quarters: (2) A.B. Beardsmore, (3) A.B. Staunton (Vice-Capt.), (4) P./O. McLean, (5) A.B. O'Connor.

Five-eighth: (6) A.B. Wild.

Half-back: (7) A.B. O'Sullivan.

Forwards: (9) S./Lt. Gourlay (Capt.), (8) Ck. Bray, (11) A.B. Aiken, (12) A.B. Kearney, (10) A.B. Sanderson, (13) Wrt./Sch. Phillips, (14) A.B. Armstrong, (15) A.B. Helsden.

Reserves: (16) P./O. Spargo, (17) A.B. Jackson, (18) A.B. Huxham, (19) Marshall.

At Manly Oval

1 p.m. Referee: Mr. B. McGREGOR.

Manly–North Districts 3rd Grade XV v. The Rest

Manly-North. Districts 3rd Grade (Royal Blue).

Full-back: L. Moore (M.L.S.).

Three-quarters: J. Forsyth (M.M.), L. Simpson (M.L.S.), R. Milligan (St.M.), J. McMillan (Spit).

Five-eighth: E. Hopkins (M.L.S.).

Half-back: G. Dallymore (M.L.S.), Capt.

Forwards: B. Bonner (St.M.), J. Perkins (N.B.), J. Sullivan (M.L.S.), J. Marvell (N.B.), K. Aubrey (Spit), R. Williamson (M.L.S.), G. Ward (Spit), D. Murray (St.M.).

The Rest.

Full-back: M. Shea (N.B.).

Three-quarters: G. Browne (Spit), G. Smith (St.M.), K. Marrott (N.B.), J. Carroll (St.M.), Capt.

Five-eighth: K. Brookes (Spit).

Half-back: J Parlett (M.L.S.).

Forwards: H. Hodges (M.L.S.), J. Hornershaw (M.L.S.), P. Rothwell (Spit), J. M. C. Bonner (M.M.), J. Woolcott (Mos.), N. Dunford (M.L.S.), R. Dutch (N.B.), D. W. Miles (M.M.).

Reserves (both teams): C. Hall (M.M.), R. Pym (M.M.), N. Brown (M.L.S.), G. Johnson (M.L.S.), D. Bellingham (Spit), M. Richardson (St.M.), D. Gordon (St.M.), J. Upham (M.L.S.).

NEW SOUTH WALES RUGBY UNION

PROGRAMME

FOR

KING'S BIRTHDAY MATCHES

Monday, June 18th, 1945

AT

NORTH SYDNEY OVAL

AND

MANLY OVAL

Price: 3d.

Printed by Tomalin & Wigmore, 545 George Street, Sydney.

The programme for the King's Birthday matches on 18th June 1945

Our try-out at Parramatta was to find a few young men from the Royal Navy establishments around Sydney to form an RN team. The next stage would be to combine the RN with the RAN for a tri-service competition. I must say that preliminary training was non-existent and other than in the matches, I never touched a rugger ball, nor did I do any fitness training. Our first appearance in public was to participate in one of the King's Birthday matches at the North Sydney Oval on Monday, 18th June 1945, a public holiday. This particular festival was under the umbrella of the NSW Rugby Union and included three other matches played at Manly Oval. Our match was between the RN and the RAN and was the first of three being played on the same pitch, one after the other. I played at wing three-quarter and we won the match 8 - 3.

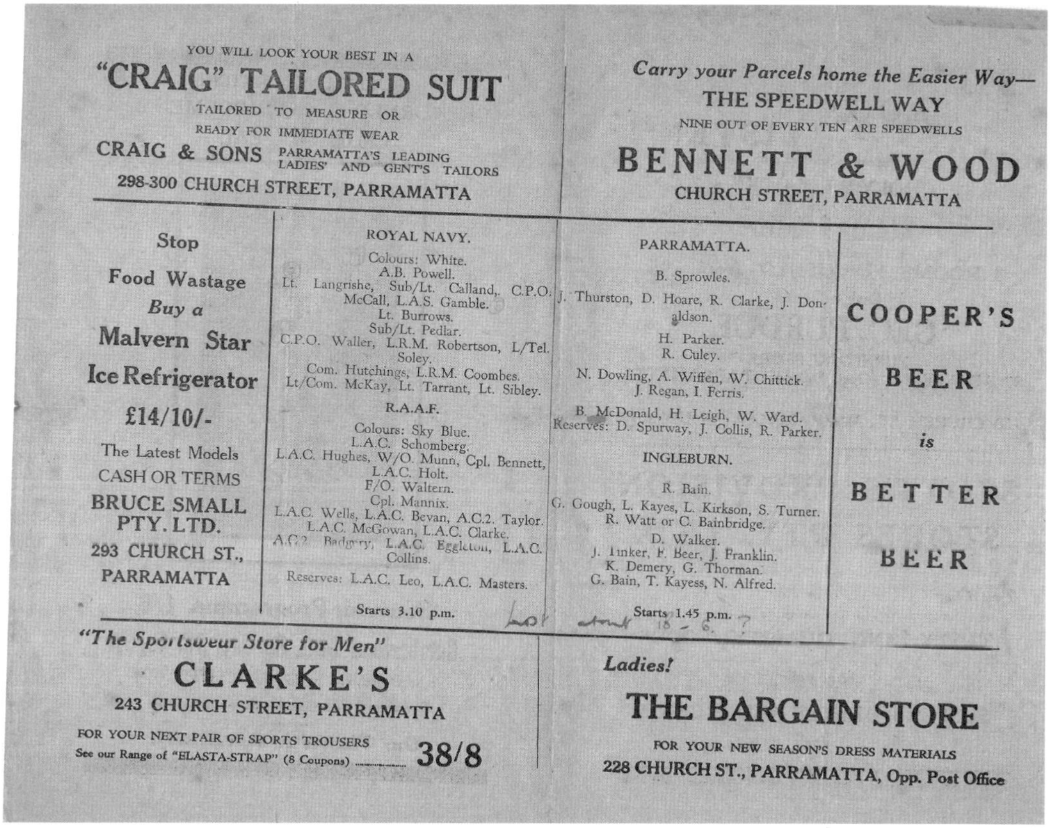

YOU WILL LOOK YOUR BEST IN A
"CRAIG" TAILORED SUIT
TAILORED TO MEASURE OR READY FOR IMMEDIATE WEAR
CRAIG & SONS PARRAMATTA'S LEADING LADIES' AND GENT'S TAILORS
298-300 CHURCH STREET, PARRAMATTA

Carry your Parcels home the Easier Way—
THE SPEEDWELL WAY
NINE OUT OF EVERY TEN ARE SPEEDWELLS
BENNETT & WOOD
CHURCH STREET, PARRAMATTA

Stop
Food Wastage
Buy a
Malvern Star
Ice Refrigerator
£14/10/-
The Latest Models
CASH OR TERMS
BRUCE SMALL PTY. LTD.
293 CHURCH ST., PARRAMATTA

ROYAL NAVY.
Colours: White.
A.B. Powell.
Lt. Langrishe, Sub/Lt. Calland,. C.P.O. McCall, L.A.S. Gamble.
Lt. Burrows.
Sub/Lt. Pedlar.
C.P.O. Waller, L.R.M. Robertson, L/Tel. Soley.
Com. Hutchings, L.R.M. Coombes.
Lt/Com. McKay, Lt. Tarrant, Lt. Sibley.

R.A.A.F.
Colours: Sky Blue.
L.A.C. Schomberg.
L.A.C. Hughes, W/O. Munn, Cpl. Bennett, L.A.C. Holt.
F/O. Waltern.
Cpl. Mannix.
L.A.C. Wells, L.A.C. Bevan, A.C.2. Taylor.
L.A.C. McGowan, L.A.C. Clarke.
A.C.2. Badgery, L.A.C. Eggleton, L.A.C. Collins.
Reserves: L.A.C. Leo, L.A.C. Masters.
Starts 3.10 p.m.

PARRAMATTA.
B. Sprowles.
J. Thurston, D. Hoare, R. Clarke, J. Donaldson.
H. Parker.
R. Culey.
N. Dowling, A. Wiffen, W. Chittick.
J. Regan, I. Ferris.
B. McDonald, H. Leigh, W. Ward.
Reserves: D. Spurway, J. Collis, R. Parker.

INGLEBURN.
R. Bain.
G. Gough, L. Kayes, L. Kirkson, S. Turner.
R. Watt or C. Bainbridge.
D. Walker.
J. Tinker, F. Beer, J. Franklin.
K. Demery, G. Thorman.
G. Bain, T. Kayess, N. Alfred.
Starts 1.45 p.m.

COOPER'S
BEER
is
BETTER
BEER

"The Sportswear Store for Men"
CLARKE'S
243 CHURCH STREET, PARRAMATTA
FOR YOUR NEXT PAIR OF SPORTS TROUSERS
See our Range of "ELASTA-STRAP" (8 Coupons) 38/8

Ladies!
THE BARGAIN STORE
FOR YOUR NEW SEASON'S DRESS MATERIALS
228 CHURCH ST., PARRAMATTA, Opp. Post Office

The match details for the 5th August 1945

My next outing was in a Royal Navy team versus the RAAF, played on the Cumberland Oval at Parramatta on Sunday, 5th August. This ground was the home of the long-established Parramatta Rugby Union Club. In researching the background of these venues I have discovered that in 1981, after the Parramatta RUFC had won the club championship for the first time ever, their fans burnt down the club; to this day it has not been rebuilt! Our match there followed a club match. We were beaten 18 - 6.

Finally, on a scorching day, Sunday, 12th August, I played wing three-quarter for a joint RN/RAN team versus a RAAF team at North Sydney Oval. We were the curtain-raiser for a rather more serious game between a Combined Services XV and a Metropolitan XV. Our match was refereed by Surgeon Commander S.R. Wallis RN. He was an engaging character who spent a lot of time talking to the players as the game proceeded, something I had never experienced before. He was a very good referee and, I believe, refereed international matches back home after the war. The match ended for me painfully; North Sydney Oval was essentially a cricket ground and the rugger field was marked out with the cricket square in the middle. The square was topped with a Bulli dressing. Although this was supposed to be winter, the sun was strong enough to harden the "grass" surface into a strong resemblance to bare concrete. To tackle or be tackled on such a surface was to experience a painful scraping. We managed to draw 8 - 8 but I could hardly walk off the field at the end of the game. One of my team-mates was Cdr. (S) Hutchings RNVR the Cdr (S) of Schofields, a charming man who took pity on me. He persuaded local Australian friends with whom he was staying the night to put me up, for which I was most grateful. It was to be several weeks before I was fully fit again and I was unable to enjoy the days of celebration to the full following the end of hostilities on 15th August, to be known thereafter as VJ-Day. I played no more rugger in Australia. One of my MS1 personnel, L.A.F. McGowan, was a very good footballer. When we were established at Schofields I readily gave him leave on Saturday afternoons to go into Sydney and play for one of the clubs in a Sydney league.

3 p.m. Referee: Mr. L. C. TOMALIN.

COMBINED SERVICES v. METROPOLITAN

COMBINED SERVICES XV. (Black and White.)	METROPOLITAN XV. (Green and Gold)
Full-back: 1—C. SCHOMBERG (R.A.A.F.).	Full-back: 1—B. CREER (E.S.).
Three-quarters: 2—C. SMITH (R.A.A.F.). D. CARRICK (Army)—3; 4—R. RANKIN (R.A.A.F.), Capt. J. NORTH (R.A.N.)—5	Three-quarters: 2—K. McDONALD (R.). J. NAPIER (P.), Vice-Capt.—3; 4—J. M. SMYTHE (E.S.). S. PROSSER (P.)—5
Five-eighth. 6—L. STAUNTON (R.A.N.).	Five-eighth. 6—J. MARRON (E.S.).
Half-back: 7—K. E. WALSH (Army).	Half-back: 7—R. MacKERRAS (G.).
Forwards: 8—A. HODGSON (Army), Vice-Capt. R. TAYLOR (R.A.A.F.)—9; 10—O. BEVAN (R.A.A.F.). E. FREEMAN (Army)—11; 12—B. JONES (Army). G. GAAL (Army)—13; 14—D. FURNESS (Army). G. STOVELL (R.N.)—15	Forwards: 8—W. SIMPSON (M.). A. BUCHAN (R.)—9; 10—R. CORNELSON (R.). H. NEWTON (E.S.), Capt.—11; 12—A. ARMSTRONG (P.). J. KEEN (E.S.)—13; 14—A. STONE (E.S.). O. HOUSTON (D.)—15; 16—E. MORGAN (M.).
RESERVES: 16—A. BURROUGHES (R.N.). E. TWEEDALE (R.A.N.)—18; K. KEEBLES (Army)—19	RESERVES: 17—J. BOLAND (P.). E. HAMILTON (St.G.)—19; 18—V. GRAY (P.). N. SHEHADIE (R.)—20; P. JOHNSON (U.). E. HEARN (P.); R. MOORE (N.S.). J. STENMARK (U.).

1.45 p.m. Referee: Surgeon-Commdr. S. R. WALLIS.

R.A.A.F. v. R.N.—R.A.N. Draw 8-8.

R.A.A.F. XV. (Sky Blue.)	R.N.-R.A.N. (Navy Blue.)
Full-back: 1—C. HOLT.	Full-back: 1—A.B. BEARDSMORE (Penguin).
Three-quarters: 2—J. HUGHES. J. MUNN—3; 4—A. HEARNE. F. PETERS—5	Three-quarters: 2—Lt. LANGRISHE (Schofield). A.B. STAUNTON (Kybra)—3; 4—S./Lt. CALLAND (Bankstown). A.B. JAMES (Penguin)—5
Five-eighth. 6—E. THACKERAY.	Five-eighth. 6—A.B. WILD (Kybra).
Half-back: 7—A. MORCOMBE (Vice-Capt.).	Half-back: 7—A./M. ABBOTT (Nowra).
Forwards: 8—A. WELLS. S. MUTCH—9; 10—W. SARGENT. W. HOLMES—11; 12—J. CLARKE. E. EGGLETON (Capt.)—13; 14—J. K. COLLINS. L. RAPPENACKER—15	Forwards: 8—Lt. KOLTS (Kybra). L./S. McNEICE (Watson)—9; 10—A.B. SANDERSON (Rushcutter). L.R.M. COOMBS (Bankstown)—11; 12—Cmdr. HUTCHINGS (Schofield). L./S. BOURKE (Watson)—13; 14—A./M. ROBERTSON (Schofield). C.P.O. WALLER (Golden Hind)—15
RESERVES: 16—K. STEWART. R. BELL—17	RESERVES: 16—A.B. O'SULLIVAN (Kybra). Sig. WOODALL (Golden Hind)—18; 17—L./S. SHAW (Rushcutter).

Players listed for both matches at the North Sydney Oval on 12th August 1945. Considering the difficulties in contacting Royal Navy players, I was surprised how many were picked.

Schofields, a busy base for carrier squadrons

During May and June Schofields had seen more visiting squadrons. *Arbiter*'s delivery of 1843 Sqn. in May has been mentioned. Kamikaze attacks during Operation *Iceberg* had severely damaged *Victorious* and *Indefatigable* so they withdrew to Sydney for repairs. *Victorious* disembarked the Corsairs of 1834 and 1836 Sqns. to Schofields on 5th June until 26th, when the ship sailed once more for Manus to rejoin the Fleet. The entire air group of *Indefatigable* flew ashore to Schofields on 5th June. Not only were there Seafires (887 and 894 Sqns.) and Fireflies (1770 Sqn.), there were also the Avengers of 820

Sqn. MS1 was detached from 706 and supported 820 with men and spare parts. All squadrons were back on board during the first week of July when the ship departed for Manus. 1772 Sqn. took the place of 1770 Sqn. in *Indefatigable*. This squadron had been the first of the Firefly squadrons to see action the previous year. Maj. V.B.G. Cheesman DSO, MBE, DSC, RM had been their CO. He was a man of great charm and charisma and was a brilliant pilot and leader; his squadron, and indeed the whole Air Group of *Indefatigable*, held him in the greatest respect. 1770 Sqn. had had an active year in *Indefatigable*, starting with Operation *Mascot* against the *Tirpitz* in July 1944 and culminating in Operation *Iceberg* to support the U.S. landings as a constituent ship of TF57, BPF. By June 1945 the CO and many of his aircrew were "time expired" and due for replacement so Maj. Cheesman handed over command of 1770 to L/C (A) D.J. Holmes RN on 22nd June and was returned to the UK with many of his aircrew. The squadron was, to all intents and purposes, disbanded and reformed with a new CO and a new group of aircrew. They remained at Schofields to work up during July and August but were moved to Maryborough at the end of the month, only to be disbanded for good on 30th September 1945.

There were no more large carriers to add to the existing carrier squadron but a second squadron began to form when *Venerable*, one of the new *Colossus* class Light Fleet Carriers, arrived in Sydney on 22nd July, disembarking the 18 Barracuda IIs of 814 Sqn. and the 21 Corsair IVs of 1851 Sqn. to Schofields. This was the first view of the Barracuda in the Pacific and, just in the nick of time, 706 Sqn. received six aircraft of the type from Bankstown, finally bringing the squadron up to its establishment of thirty-six aircraft. *Venerable's* squadrons re-embarked after three weeks ashore and the ship proceeded to Hong Kong (the war by then being over) to cover the re-occupation of the colony, accompanying *Indomitable*. Three other ships of the *Colossus* class also arrived in Sydney. They were *Colossus* herself, *Vengeance* and *Glory*. All were equipped with a squadron of Barracuda II and a squadron of Corsair IV. *Glory* disembarked 837 Sqn. (Barracuda II) and 1831 Sqn. (Corsair IV) to Schofields on 16th August, only to re-embark on 1st September. While all this was going on, *Colossus* and *Venerable* had landed their Air Groups at Jervis Bay. Together these four Light Fleets were to form the 11th Aircraft Carrier Squadron. The cessation of hostilities changed their role. For the next few months they were used to cover the re-occupation of British colonies in the Far East, to accept the surrender of Japanese troops and help in the rescue of POWs from Japan.

We had heard all about the two atomic bombs, news of which had convinced us that our lives were no longer in mortal danger, so during the days after 15th August there was a profound sense of relief throughout the Navy, the Fleet Air Arm and the good citizens of Sydney. People in the United Kingdom today seem to have more sympathy for the Japanese than for the servicemen who ground their way across the Pacific on behalf of their country. To end the war then undoubtedly saved thousands, even millions, of lives which would have been lost if Operation *Olympic*, the invasion of Japan, had taken place. Flying activity at Schofields immediately came to a screeching halt. Cdr (A) G. Kenton RNVR, Commanding Officer of HMS *Nabthorpe*, MONAB3 of RNAS Schofields, cleared lower deck and we all paraded to hear him tell us officially that the war was over. Normal duties were suspended and most of the personnel on the Station took the opportunity to clear off to their friends in Sydney or to join the crowds in Martin Place, or both.

During my association with 706 Sqn. I had got to know the squadron staff very well. During July a group of experienced pilots and observers of lieutenant rank had been flown out to Australia to await postings as squadron commanders. For the time being they were attached to 706 Sqn. for flying practice. One of them was Lt (A) G. McC. Rutherford DSC RNVR whom I had met at Fearn nearly two years before when he was with 823 Sqn. flying Barracudas. At that time junior Station officers shared the wardroom in A Camp with the visiting squadrons and many friendships were formed. "Mac" Rutherford had taken me up on one occasion in the observer's seat in a Barracuda. He was to do the same one day with me in the observer's seat in an Avenger of 706 Sqn. when the squadron was practising formation flying. I enjoyed a spectacular view of Sydney. Sometime during August, and without any warning, I was informed that I was appointed to 706 Sqn. as their Air Engineer Officer, vacating my post in charge of MS1. Again in this case there was no written order for the appointment. I was to hand over responsibility for MS1 to Johnnie Clemes, who adopted my men, trailers and stores and absorbed them into his operation of MS2. Again, there was no stocktake and handover was free of paperwork! Within a few days of VJ-Day – it was either immediately before or immediately afterwards, - "Tommy" Thom asked me if I would like to have a few flying lessons in one of the Station Flight Tiger Moths. I leapt at the chance. In the few days around VJ-Day Schofields was a busy place. There were around a hundred aircraft in daily use, of which 706 Sqn. mustered thirty six; the crew room was full of eager young men looking for flying practice.

De Havilland Tiger Moth R5130. This aircraft is photographed over England. The Tigers we were issued with in Australia were Australian-made and were borrowed from the RAAF. They were painted "Training yellow" all over and carried RAAF numbers. The aircraft I flew in were A17-721 at Schofields and A17-722 at Maryborough.

Coffee after lunch on the balcony of the Staff Wardroom, BPF Headquarters, North Shore, Sydney, August 1945. Across the harbour can be seen the public park known as The Domain. Today you would see the Sydney Opera House. Left to Right: Lt Langrishe, two unknown Lt AEOs, Lt Robertson, Lt (E/P) G.F. Hawkes RN, Staff Test Pilot.

Squadron Air Engineer Officer

Whose decision it was, I do not know, but it was evidently decided that while any of our carriers remained in the Pacific, they would continue to be ready for action, their air groups fully manned. In which case there needed to be a source of replacement aircrew, not only to replace casualties but also to enable long-service men who were due for home leave and demobilisation to be released. Following VE-Day, the British Government had introduced a scheme of "Age and Service Release", based, as its title implies, on one's age and length of service. This was immediately applied to British servicemen in Australia, India, Ceylon, Burma and the south west Pacific, and the relevant fleets. 706 Sqn. appeared to have a continuing role but, for whatever reason, it was immediately to be elsewhere. On 28th August a large formation of the squadron's aircraft set off for Maryborough, Queensland, the temporary home of MONAB6, which was to be our next base for an unknown period. Perhaps we were surprised that all thirty-six aircraft arrived safely, with no precautionary landings en route. All our kit, some ground equipment and a selection of ground crew made the trip in a Dakota supplied by the RAAF. Most of the hundred-odd ground crew went by train, a long and tedious journey. I flew in the observer's position in the middle of an Avenger flown by Tommy Thom. Being a Sydney-sider Australian, he was fully aware of the whereabouts of the various Australian commercial medium-wave radio stations up and down the coast. Several of our older Avengers were still fitted with an American multi-wave-length radio set so once free of Schofields Air Traffic Control Tommy tuned in to the strongest signal on medium wave. As we progressed up the coast he kept us happy with the popular music of the day from one station after another. We had, I am sure, already changed the squadron CO for the last time and we now had Lt Cdr (A) C.A. Fraser RN as our CO.

Angus Fraser was an Observer of great experience, having been one of that band of "A-boys" who had joined the Navy as short-service commission aviators in the late 1930s. They had been stalwarts throughout the war, the late Lt Cdr (A) E. Esmonde VC, DSO, RN of 825 Sqn. being a notable example. Angus was a tall red-haired, red-bearded man with a good sense of humour, natural authority and dignity. In the closed community of a squadron he was at one and the same time "one of us" and our leader. Personally, I enjoyed his command and the next few months were happy ones for me. I now really had to get to know the squadron's "Chiefy", the Flight Sergeant I have already mentioned. I had been in regular touch with him while MS1 was working with the squadron's Avengers but I now had to take over some of the administrative matters he had dealt with up to now, to act as the spokesman for the engineering department with the squadron commander and to be their divisional officer. I think I did the job without alienating such a valuable member of the squadron. As had been the case when I was chosen to run ARS at Fearn and when I took command of MS1, no advice or instructions were given on how to manage the units, or, indeed, details of what my function should be. I felt I was in a sensitive position and this feeling persisted throughout the next seven months. I made no notes about how we coped with the reduced requirement of maintenance personnel when we reduced our aircraft numbers. The only appropriate memento is a photograph of the squadron taken later in the year at Nowra. There were fourteen officers, of whom five were the squadron CO and his staff, an electrical sub-lieutenant and the rest attached aircrew; there were 116 Chiefs, Petty Officers and ratings. How many of these men might not be technicians I do not remember, but presumably the whole range of technicians would include radio and radar mechanics and ordnance ratings. This number would appear to be more than enough for twelve aircraft. It is puzzling that, for the photograph, we could not muster enough TBR or Firefly crews, nor single-seater pilots, to man our twelve aircraft.

The aerodrome at RNAS Maryborough was on the edge of the small town of that name, 180 miles north of Brisbane. It was an RAAF Station, housing a radio school, but was also designated as Maryborough's civil airport and was in occasional use by a small commercial airline. It had three runways of tarmac construction in the usual triangle, in flat surroundings, but there was at that time no resident RAAF flying unit. MONAB6, HMS *Nabstock*, was a lodger unit and had been in residence only since June. Also temporarily attached was "A" Flight of 1701 Squadron. This was an Air-Sea Rescue squadron commanded by Lt (A) L.F. Plant RNVR, an experienced Walrus pilot. "B" Flight was resident at Ponam (MONAB4) in the Admiralty Islands for the same duties. Leonard Plant had chosen to stay at Maryborough while his Senior Pilot took charge of "B" Flight. Each flight had two Supermarine Sea Otters. Unfortunately for the Navy at Maryborough, the RAAF retained all catering for the whole camp and provided all facilities for the officers and men of both services. Accommodation for our squadron staff officers was in two-bed cabins of very modest size. The Air Force officers' mess hut was an unfriendly place and the food only just acceptable.

Having arrived in Queensland with all these aircraft we awaited the next stage in the knowledge that such a number must be unnecessary for the immediate future. Within a few days we were to lose a large number of our replacement aircrew, most of whom were posted home. We saw off a party of them on the train from Maryborough, eventually to be deposited in Sydney. Next, we had to reduce our holding of aircraft to two of each type, TAMY1 at Archerfield, Brisbane, taking our surplus. With new aircraft still going through final assembly and testing, and the depositing of large numbers of aircraft from disbanded squadrons, Archerfield was awash with aircraft. In collaboration with Cdr (S) at Maryborough and his equivalent at Archerfield, I was able to negotiate a deal by which we were to end up with two each of the very latest zero- or low-hours aircraft. We would then have FG-1D Corsair IV, F6F-5 Hellcat II and TBM-3 Avenger III, all finished in midnight blue and with "zero-length" rocket-projectile mounts. The Avengers had an ASH radar pod under the outer starboard wing – we all called it an "ash can". Seafire LIII, Firefly FR1 and Barracuda II were all brand-new, the Fireflies also having an ASH radar pod under the engine. I made one flight to Archerfield in a Barracuda and found the airfield there an extraordinary sight. There were Naval aircraft almost wing to wing all round the airfield.

Supermarine Sea Otter ASR-I JN185. Type used by 1701 Sqn. "A" Flight at Maryborough. The Sea Otter was the last of a line of small single-engined Supermarine flying-boats going back to the "A.D. Boat" of 1916. The current role was Air-Sea rescue.

We were very proud of our new equipment. The shiny blue American aircraft spurred our maintenance ratings to make our rough British camouflage look the same. The Seafires were small enough to be manageable and we bought and used many cans of car polish to improve their appearance. This was, of course, highly irregular but we considered it unlikely that the aircraft would ever see action so – what the hell! We may even have found a few more miles per hour in top speed. MONAB6 had also been issued with two Tiger Moths and Tommy Thom continued to instruct me how to fly and I went through the routines of climbing, gliding, stalls, turns, course steering, medium and steep turns and loops. We made one longer flight up the coast to RAAF Bundaberg to practise map-reading, and for the purpose of scrounging an aero-engine part for the Double Cyclone engine in one of our Avengers. The same basic engine was fitted in the RAAF B-25 Mitchell bombers based there.

Our daily routine was never very demanding and usually finished at 16.00. We found a café just opposite the main gate and the four of us staff went across the road most afternoons for a high tea. Invariably this was an egg dish done in a variety of ways and we usually ended up having eaten two two-egg dishes. We made no efforts to get to know the RAAF officers in the mess nor the people of Maryborough. There were two cinemas in the town, within walking distance of the airfield, and a camp cinema. All of them changed their programmes twice a week and the camp cinema had a special programme on Sunday evenings. For want of anything else to do, in one week I went to the "flicks" seven evenings in a row! Before the walk back to camp we would find a milk bar for a cold drink or ice cream. One evening we attended a "hop" in a local community hall, probably a fund-raising event in aid of the Returned Servicemen's League, or RSL.

On our arrival at Maryborough at the end of August we had been surprised to see a collection of Corsairs with the identity of *Victorious* scattered round the airfield. For some reason 1834 and 1836 Sqns. had shared their disembarkation from the carrier between Maryborough and Bankstown as she steamed south on her way to Sydney. Those left at Maryborough were flown to Archerfield for destruction. This was a melancholy business forced upon us by the terms of Lend Lease and was not concluded until as late as March 1946.

There were only two technical incidents at Maryborough which spring to mind. Firstly, we lost an Avenger which was force-landed in swampy forest near the River Mary, a couple of miles from the airfield. The crew of two escaped the crash unharmed. Presumably we raised an A25 accident report form and saw to the recovery of the aircraft, but that memory has disappeared. The second event concerned one of our old Avengers. The port elevator was defective in some way and I considered that it should be replaced. A new one appeared from somewhere, but when offered up to the tailplane it would not fit; the relative positions of the hinges did not match by as much as half-an-inch. Shades of the earlier hydraulic system problem, it seemed to me that the situation had arisen because the assembly jigs at different factories had not been checked for compatibility – Grumman at New York and General Motors at Trenton, New Jersey. This was quite a shock to me because at home we had long since passed and reconciled the problems of sub-contracted parts and assemblies. I understood that, at home, it was the responsibility of the parent company to produce all the necessary jigs and tools so that the parts produced were identical and interchangeable. I always found this to be the case. With this elevator I also understood the reason for different colours appearing in photographs of US aircraft in squadron service on carriers. Whereas all British replacement assemblies would be issued in a light green primer, US factories finished all theirs in black. Some historians have tried to relate a different tail colour, say, to a squadron or ship identity. Usually this would not have been the case; there just was not the time to spray the part in its matching colour.

To be in Queensland in the southern spring was delightful. Temperatures were similar to high summer in England and there was plenty of sun. Thunderstorms became more frequent as the days passed and they could be very dramatic. We only had covers for engines and cockpits so, as we had found earlier, metal surfaces were hot, even dangerous, to the touch. The coastal strip is very productive; Maryborough is known for its production of sugar cane, much of which at that time was brought down to a jetty at Pialba, on Hervey Bay, by narrow-gauge railway, for shipment to sugar factories elsewhere. During October I was allowed a weekend's leave and went down to Brisbane and on to the south-east coastal suburb of Wynnum. My hosts were Mr. and Mrs. Stone. He was an academic in Brisbane University and was the brother of one of the Edinburgh University circle of wives, whose husband was the Professor of Botany and Superintendent of the Edinburgh Royal Botanic Gardens. The Australian connections lived in what I had come to appreciate was a typical Queensland bungalow, of which I had seen many in Maryborough. It was built on stilts, of wooden construction and cladding, with a corrugated iron roof. There was no piped hot water system. In the bathroom was a wood-fired boiler which was flashed up when a bath was wanted; wood was stacked under the bungalow. It appeared primitive but was extremely practical and efficient, particularly bearing in mind the average air temperatures. Their son had a leather-working business in the city, and I was taken to see the factory. I returned to Maryborough from Archerfield in a Beech Expediter of 724 Sqn. on its scheduled flight from Sydney.

Supermarine Seafire LIII on assembly line at Archerfield, 1945. These were brand-new aircraft, two of which were issued to 706 Sqn. at Maryborough.

By the middle of October it was apparent that there was to be an extensive but orderly reduction in all Royal Navy establishments in Australia as the British Pacific Fleet and the Fleet Train shrank. 706 Sqn. was ordered to transfer to Nowra, a long way from Maryborough. Our twelve aircraft flew south on 24th October, the movement once more being assisted by a RAAF Dakota and the railway system. Queensland's State railway was constructed to a narrower gauge than 4 ft. 8 ½ ins. so all rolling stock was smaller and the trains ran more slowly than in NSW. There was an interchange of gauges in Brisbane. I was given the rear seat in a Firefly FR1 flown by S/Lt (A) Demaine RNVR, who chose to fly low enough to give us a good close-up view of the coast. In fact, we were close enough to experience several near misses of seabirds. It was very exhilarating. We had to land about halfway to Nowra for fuel, which I believe was at a RAAF airfield at Coffs Harbour. Our low flight took us past Sydney Heads at nought feet so, to my disappointment, we saw nothing of the harbour or of the city. The total distance was over 900 miles, which, I noted, was covered in 3 hrs. 15 mins. I find this time questionable as it represents a cruising speed of 281mph. I suspect that the actual time was nearer 3 hrs. 45 min.

From that date onwards Fleet Air Arm activity at Maryborough was gradually run down and those on relevant Age & Service numbers were sent home. The nucleus of *HMS Nabstock* was withdrawn on 19th November and took over the administration of RNAS Schofields in this name, whereupon *HMS Nabthorpe* went out of commission, releasing a further batch of personnel for demobilisation at home.

I arrived at Nowra to find most of my old friends of MONAB1 were still there. Some building work had been done on the camp and the airfield, but no hangars had been built for squadron use. The wardroom mess and the officers' accommodation huts were in the same place as before. 706 Sqn. was assigned the dispersal area at the south-east corner of the airfield, close to the east end of the east-west runway, with a line of huts for squadron use. I was allocated a cabin in the wardroom hutment. There had been developments in the training facilities available for visiting squadrons. There was now a range for air-to-ground bombing, gunnery and rocket firing down on the coast, and 723 Sqn. had Martinets for air-to-air gunnery practice. They also had a small number of Corsair IV for towing banner targets for ships' anti-aircraft practice. Looking in the main station workshop hangar one day I saw an ex-Royal New Zealand Air Force Avenger which was being converted to target tug. I never saw it flying. To have such facilities available there and then made life for 706 Sqn. so much more interesting and effective for air and ground crews alike and meaningful training programmes could be devised for the new aircrew. They all revelled in it. For most pilots, and probably for our ordnance mechanics, this was their first experience of rocket firing, all three American types being equipped for this purpose. The latest Avenger III had the Sperry automatic pilot, equipment which was quite new to us. It might seem unnecessary in such a relatively short-range aircraft, but it reflects the standard practice in the USN of having no navigator/observer in a TBR crew; the pilot was trained to navigate and had a map board under the instrument panel which slid out for use. To have an automatic pilot was bliss. In our use it meant that, out on a "nav.ex" and once on course, the Avenger pilot switched on the auto-pilot and, literally, put his feet up, leaving his observer to do all the work. It was also known that the pilot would light up a cigarette and open one side of the sliding hood to ventilate the cockpit. "All mod cons".!

Both our staff observer Lt Clayton and the trainee observers on strength learned how to use the ASH radar fitted to the Avenger and the Firefly FR1. At this time of the year it was important to know if there were thunderstorms about because they could be of tropical size and intensity. Before sending Avengers, Barracudas or Fireflies out on a nav.ex, the Senior Observer would consult the Station Met. Office several hours before take-off was due. Our met. men were now fully attuned to the weather pattern of the Australian continent. The characteristics were quite different to what we had come to expect in the northern hemisphere and on the edge of the Atlantic Ocean. Making use of all the Air Force met. offices across the continent, the large weather systems over Australia were tracked by the minute. We could be told exactly how fast a centre of low pressure or a front was moving in from the west and when the latter would arrive over Nowra. "Exactly" is no exaggeration. Often we experienced a downpour within a few minutes of the scheduled return of our aircraft, when the rain could be seen bouncing nearly two feet off the tarmac.

Avenger TBF-1 NZ2544; the type was used by the RNZAF in the battles for the Solomon Islands in 1944. A small number were turned over to the Fleet Air Arm in 1945 to be converted to target tugs. One was in the hangar at Nowra in October 1945.

As an example of how long it takes to build up the infrastructure of an active airfield, it was not until about August 1945 that the supply of motor transport caught up with requirements. A load of Canadian-built Chevrolet and International trucks from 15 cwt. to 3 ton had been shipped and distributed. When we arrived at Nowra we were told by the MT department to go and choose what we wanted from the line-up in their vehicle park available for visiting squadrons. I think we took a couple of each of those sizes. I also found an American Indian 600 c.c. side-valve vee-twin-cylinder motorbike which I was allowed to take. This proved most useful – once I had become used to the differences to a British machine. I could almost feel I was an American highways cop! The use of Service transport for social activities was not forbidden but had to be justified. In such an isolated community this facility was appreciated. How else could one get down to the beach or bars at Jervis Bay? Public transport was only available to and from Nowra town by a public bus service, using the direct road from the north end of the camp – a dirt road.

There had been a number of squadron transfers through Nowra during the year as carriers came and went from the BPF. But as the ships retired from the north Pacific for good in the last few weeks of August squadrons flew all their aircraft ashore as they disbanded or were sent home without aircraft for disbandment in the United Kingdom. At the end of that month fifteen Avengers of 848 Sqn. and thirty-six Corsairs of 1841 and 1842 Sqns. came ashore to Nowra from *Formidable*. Their aircrew returned to the UK in *Victorious* for formal disbandment. At the same time 828 Sqn. disembarked from *Implacable*. Lt Cdr (A) R.E. Bradshaw DSC** RN took over command of what was in effect a new squadron after the withdrawal of time expired aircrew and re equipment with twelve new Avenger III. They used all the facilities of Nowra for working-up before moving to Schofields on 27th November. On return from the Pacific, *Indefatigable* also disembarked her Avenger squadron No. 820 to Nowra on 18th September. They had also reduced to twelve aircraft. They re-embarked on 23rd November for the goodwill trip to New Zealand and Melbourne and never returned to Nowra. The last of the Air Groups on the Fleet Carriers to use Nowra was from *Indomitable*. 857 Sqn. Avengers, 1839 and 1844 Sqns.' Hellcats all arrived on 11th October. Leaving their aircraft behind, crews returned to their ship which sailed for the UK where the squadrons were officially disbanded. 828 Sqn.'s Avenger III were lined up on the large dispersal area near the control tower where I used to walk past on most days. I was appalled to see a large puddle of oil under the engine of every aircraft – these were machines with probably no more than thirty flying hours. I was reminded of a remark by one of the instructors at the Rolls-Royce engine school. The argument about liquid- and air-cooled engines never went away and our instructor disparagingly described Bristol engines as oil-cooled. This could be applied to all radial engines and to cope with this generous use of engine oil all aircraft so fitted had a much larger oil tank than aircraft with liquid-cooled engines. I recall that as part of the start-up procedure one was expected to "pull through" a radial engine by hand for at least one revolution in order to clear the lower cylinders of the oil which had dribbled down the cylinder walls from the crankcase. If you disregarded this instruction you could end up with a hydraulic lock and a damaged engine; with so much oil about in unexpected places, starting these large radials was usually accompanied by a cloud of oily exhaust smoke. There is an air-to-ground photograph of an impressive line-up at Nowra taken a short time before the arrival of 706 Sqn. on 24th October. It shows thirty-six Corsairs, twelve Hellcats and nineteen Avengers.

This picture can be dated as September 1945. The end of the war in August triggered a mass disbandment of squadrons from the Fleet carriers. Some disembarked their Corsairs and Hellcats to Nowra and went home, taking their aircrew with them to disband in the UK. In this period as many as seventy-two Corsairs were delivered to Nowra by 1841 and 1842, while 1843 and 1845 had been spare squadrons based at Nowra. The Hellcats were the remains of thirty-six aircraft of 1839 and 1844 Sqns. Six Avenger squadrons (except from HMS Victorious) left aircraft at Nowra but not at the same time. At one time in September there could have been over sixty Avengers on the tarmac; note the dark patches where aircraft had been standing.

So by the time we arrived at Nowra much of the work relating to the shrinking of the fleet had been done. After a fortnight at Schofields the *Glory* Air Group had re-embarked on their carrier for a few days of flying practice then, on 11th September, they disembarked to Jervis Bay which had been operated by MONAB5 since May. In preparation for the closure of Jervis Bay these two squadrons, 837 (Barracuda II) and 1831 (Corsair IV), moved up to Nowra on 29th October and 837 exchanged its Barracudas for a set of brand-new Firefly FR.1.

Around this time I watched a remarkable performance by the 1831 Sqn. Corsairs. They had been up for a squadron "Balbo"[8] with all twenty-four aircraft and came back to do a stream landing on runway 26 which was over 2,000 yards long. With no mishaps, by the time the leader was turning off the end of the runway, more than half the squadron were already on the runway behind him. This air group stayed at Nowra until the middle of January 1946 when they re-embarked on *Glory* for a month. So for the rest of the year, Nowra was a busy airfield with much to enjoy. We had our disasters too; one of our Corsair pilots crashed into the bush between Nowra and Jervis Bay, killing himself and destroying his aircraft. Then a pilot was drowned in the designated swimming pool in the Shoalhaven river, becoming entangled in water weed. It seemed so pointless for these young men to lose their lives two months after the war was over. We gave them proper funerals.

By November 1945 further change was in the air. Groups of older men departed regularly for Sydney to join the troopships for home. This applied as much to Jervis Bay as it did to Nowra. The remaining personnel of MONAB1 and MONAB5 were combined under the umbrella of MONAB5 and Nowra became *HMS Nabswick* on 15th November following the closure of Jervis Bay. It is untrue to say that one took over the other.

There were occasional messages from the Admiralty about peacetime prospects in the Royal Navy for officers and ratings but, certainly in my case, it was difficult to make a decision without seeing what was going on at home. A small number of men chose to be demobilised in Australia, having found either a job or a wife, or both. We made much of our own amusement. A wardroom dance was organised; Nowra's resident officers had established friendships with families in the town and could produce partners. We visitors had to turn to the bevy of young women staying in the hotels and boarding houses in Jervis Bay village. We probably had to collect and return them by lorry, but this added to the excitement of the evening. One weekend three of us 706 Sqn. staff took the squadron's 15 cwt. truck for a drive and picnic in the bush upstream from Nowra and into the Kangaroo Valley. As we drove into the bush on a dirt road it soon became apparent that it would be very easy to get lost. As far as the eye could see there was eucalyptus forest and no sign or sound of human habitation. I was relieved when we turned round and retraced our tracks successfully.

I never understood how the aircraft abandoned at Nowra, Schofields and Maryborough by the squadrons eventually made their way to Bankstown or Archerfield, the only two of our airfields which actually stored or destroyed aircraft. They must have been flown there, but who made sure that they were in flying condition and who flew them away? I can only presume that MONAB air mechanics had to do the job of flight preparation and fill in the F700, but there must have been instances of unserviceability, followed by the rigmarole of obtaining spares. It is probable that the teams of test pilots at Bankstown and Archerfield, many of whom were still on loan from the RAAF, were by September 1945 testing very few new aircraft, so could be used for ferrying. They would be flown out in the Expediters. Looking back, I can remember nothing of how damaged aircraft were transported to the two repair Stations. I think we had to call on the RAAF to lend us such vehicles as the "Queen Mary" articulated low-loader trucks. These were certainly needed to carry aircraft down to Woolloomooloo for loading on the carriers for ditching - and they must have been used to take the aircraft the other way. The same applied at Archerfield and transport to and from Brisbane docks.

No sooner were we installed at Nowra than I took the leave I had been hoping for at Maryborough – although not to the same people. My mother had put me in touch with an old school friend of hers

8 Named after General Balbo, who led a formation of Italian flying boats from Italy to America and back again in 1933, this referred to a large formation of aircraft.

who had married an Australian Army doctor in the Great War, Dr. Southey, who had served on the Western Front. Mother had not seen her since she shipped to Australia in 1919. But, as was a common habit in those days, friends kept in touch by letter, wherever they were in the world. Surprisingly, the Empire mail system did not stop during our war and, as soon as I disclosed that I was going to New South Wales, my mother wrote to her friend. Dr. Southey had taken on the job of GP in the little town of Jerilderie, in the agricultural area of the Riverina in the middle south-west of New South Wales. It was Billabong Creek country, at one time the home of that famous criminal character Ned Kelly. Jerilderie was no larger than a large English village, with just two straight streets at an angle to each other, but it had a railway station and a night train from Sydney, 900 miles away. This was to be an experience! On 2nd November I went up to Sydney on the Expediter, via Jervis Bay to Mascot, then to Sydney Central for my train. Instead of corridor coaches on some long-distance trains, NSW State Railways had an arrangement of individual compartments, each with its "en suite" washroom and toilet entered by a door in one corner. There were no refreshment facilities. My fellow passengers were all RAAF men, going either on leave to places en route or to the Liberator heavy bomber OTU at Tocumwal, forty miles beyond Jerilderie. These chaps knew their way around. As soon as the train pulled out, two of them emptied the luggage racks and proceeded to climb in and go to sleep. Somehow I slept for several hours as we climbed over the Great Dividing Range and down the other side. I was advised to be on my toes for the breakfast stop at Junee junction; time and space were limited, but the "tucker" was said to be good, so as soon as the train stopped, there was a mad rush for the café. To my consternation, as we breakfasted, the train had been moved and had shed carriages, some to go on down the main line to Albury and Melbourne, the others on the branch line to points further west, where we were going. My carriage was in quite a different place beside the platform. But I found it. We had another 120 miles to go. From time to time we saw families of kangaroos, many galah parrots and emus beside the railway. By lunchtime we were deposited at Jerilderie. This was the back of beyond, inhabited by charming people.

The Southey family's house was out of town, a typical habitation in the hot open flat lands of the Riverina. It was a clapboard bungalow with a corrugated iron roof; internally, the only door in the house was on the bathroom, otherwise all rooms had light curtains for privacy. The hot water system was just as I had experienced in Queensland. Looking across country the most outstanding buildings were the huge shearing sheds, probably each being two hundred feet long, looking like aircraft hangars. Windmill towers were dotted about beside every house in the country to pump water. My hosts were extremely hospitable, doing all they could to keep me amused. One day I went out to a small station where they farmed sheep and I assisted with the annual job of removing the lambs' tails. Although it was only springtime it was very hot and the day had started at some unusual hour, so we stopped for breakfast at 8 o'clock – a meal based on lamb cutlets and other sheep meat, enough for a good lunch. I was given a short break and took a walk in a small open wood of gum trees, surprised to find myself frequently mobbed by flocks of budgerigars. Although it was spring, the air temperature was already higher than an English summer and I admired the way the locals coped with the heat. I was told that some families had a flat in Melbourne to which members of the family would retire for cooler summer temperatures near the sea. The air conditioners we expect today were non-existent. The time passed quickly, and I returned to Sydney on 15th November, then to Nowra by Expediter the following day. Christmas was not far off and on that day I was invited by Mr. and Mrs. Davis and their family in Nowra to a typical Aussie Christmas dinner – in the middle of the day, on the lawn in brilliant sunshine. All the food was cold – chicken salad with cold Christmas pudding and ice cream, washed down with cooled Australian white wine.

Airspeed Oxford V Instrument Trainer EB490, as used by 702 Sqn at Schofields. The Fleet Air Arm received seventy-eight Oxfords in the series PH

Approaching Christmas we had heard that our squadron was to move again. This took place in the middle of January by air and by train when we returned to Schofields, as it turned out, for the last time. We found ourselves side by side with 702 Sqn., a new second-line unit which had arrived just after VJ-Day. It was an instrument flying training and checking unit and was equipped with a Harvard, an Oxford and a RAAF Tiger Moth. We occupied the same dispersal as before. The Seafires and Fireflies from *Indefatigable* were there until the end of January 1946 when they re-embarked for passage home. 828 Sqn. Avengers and 801 Sqn. Seafires from *Implacable* re-embarked from Schofields during January, being joined by the Firefly NF-1s of 1790 Sqn. Both ships were joined by *Glory* for a goodwill visit to Melbourne in January. I have taken these movements from reference books, but there is inconsistency in the records. Evidently *Indefatigable* did not go straight home from Sydney after embarking her squadrons. There is an excellent air photograph of all three carriers alongside in Melbourne docks, taken in January 1946, published in Fly Navy No. 18, Summer 1984. They were there for a goodwill visit and had their aircraft lined up along the flight deck.

Back in Schofields we received the final member of 706 Sqn.'s staff, an Army Liaison Officer (or ALO). Apart from telling us what we should know about army co-operation, Capt. John Parker, Royal Artillery, introduced our fighter pilots to artillery spotting. He knew where the local gunnery school and range were and we struck up collaboration with their staff. It was an added interest for our pilots, spotting for 25-pounder guns firing live rounds. Also we invited some of their officers to come and see our aircraft and have lunch with us in the wardroom. In the same Australian Army training area was a tank school, residents being the remains of the Australian tank division which had been in action in the Middle East and then in the Solomons. We were invited to go over and be shown how to drive a tank. For training purposes they were still using US General Grant tanks, lumbering things which seemed to go no faster than a walking pace. It was fun knocking over eucalyptus saplings. We still had work to do with Bankstown. One day we had to take an old Avenger over for scrapping – it must have been left behind by one of the squadrons. I thought we had done everything according to the rules until I received a note from Bankstown Air Stores a few days later. Where was the aircraft clock? I was asked. As the aircraft was to be broken up for ditching within a day or two, I thought this was a bit too officious. I managed to extricate myself from trouble, but it was a lesson that one should never take the Naval Supply organisation for granted. My old friend Johnnie Clemes was still at Schofields but I cannot remember what had happened to our two MS units, their staff and stores.

By the kindness of their hearts NSW State Railways continued to allow us to travel free on their suburban network. With a more relaxed peacetime routine, we could go into town quite frequently, often in John Parker's Jeep. This vehicle had a story to tell. It was adorned with a number of unit and Army identities, most of which were spurious. Parker had acquired the vehicle in what he admitted was an underhand manner and it was to all intents and purposes his very own. With its evident war record, Parker filled up with petrol at any Service pump which was convenient. He wanted to "demobilise" it (in some Service jargon the word was "liberate"), but I never knew if he was successful. Our CO Angus Fraser also obtained a Jeep at Schofields, officially, which will crop up again later. I made contact with my friends in Sydney and I was away from the camp almost every weekend.

Hugh in McQuarrie Street, Sydney in January 1946

We were told about the scheme for sending food parcels home and we made full use of it. Stores like David Jones were participants and I used to try to send my parents a weekly parcel, sometimes adding one for my sister who had recently been demobilised from service as a Naval VAD. Sydney had changed since the end of the war with the revival of all sorts of entertainments. In the thirties Australia had been very active in speedway racing and many of their riders came over to England for the season here. English riders would go to Australia for the season down-under. Pre-war there had been international competitions, which were called Tests. Speedway racing had been re-started at the Sydney Sports Ground, an oval almost next door to the Sydney Showground and Sydney Cricket Ground, the famous SCG. There were two rival organisations, but only in name; most of the officials were the same, some with different hats. One group had meetings on Friday evenings, with a "Championship" on Monday 28th January, the other group on Wednesdays. Traditionally, racing was in a club format, but due to the "ad hoc" nature of the sport at that time, the organisers put the riders into geographical groups to create competition. I have nine programmes for meetings between 23rd January and 8th March 1946. Normally racing was to the standard of four riders and four laps at a time, all starting together. However, they introduced handicap racing when there were five riders on the track, some in front of the gate, some behind. This was most spectacular and roused the crowd to frenzy as all five would race the last lap together. There were also sidecar races, the machines being rigged with a permanent slant towards the centre of the track. This was also most exciting. Normally we went in the CO's Jeep. All went well until one evening it was found by some light-fingered gentleman who stole it while we were watching the racing. I should explain that the Jeep was vulnerable to this sort of thing because it had no ignition key. Ignition was controlled by an ordinary side-to-side switch on the dashboard so there was no magic in starting it. Fortunately, Angus Fraser had driven us in that evening so he had some explaining to do the following morning. We found our way to Sydney Central station by public transport and caught the train back to Schofields.

Facilities for visiting troops were still being provided by the people of Sydney in 1946. There was an officers' hostel and reception desk on Macquarie Street where one could get B & B. and advice about places to stay. A large hospitality building for ratings had been built from scratch in The Domain, overlooking Woolloomooloo. This was hardly complete when the war ended. Clemes and I visited the department responsible for repatriation on several occasions during February 1946 and our Age & Service numbers came up for return home the following month. We were to travel in MV *Stirling Castle*, one of the large ocean liners which had serviced the BPF from Liverpool since early 1945 but was now running back and forth from Southampton, taking troops in the other direction. She was of 25,500 gross tonnage, diesel-driven, and had been completed in 1936 so she was one of the more modern ships in the trooping fleet. There were some emotional thanks and "good-byes" to my friends in Sydney, with requests to come and see me and my parents if they came to the UK. There were also farewell parties to be experienced in the Schofields wardroom and "good-byes" to be exchanged with station colleagues and some emotional farewells to 706 Sqn. friends with whom I had served for the past year. Addresses were exchanged, most of them never to be used after one or two Christmases. The one person with whom I maintained contact was Johnnie Clemes; I dined with him and his wife only a few months before his death a short time ago in his home in Herefordshire near where I now live.

Finally I said farewell to Sydney on or about the 13th March 1946 as *Stirling Castle* let go of the ocean quay in Darling Harbour – a traditional occasion with streamers from the ship to the quayside and cheering from ship and shore, a contrast to the low-key arrival of the ship in Southampton a month later.

Fleet Air Arm activities at Schofields lasted for only another few weeks. 706 Sqn. disbanded on 31st May 1946 and RNAS Schofields finally closed down during the following month.

Journey to Civvy Street.

The route home was to be round the south coast of Australia, through the Great Australian Bight to Fremantle, across the Indian Ocean to Aden, through the Suez Canal, stopping at Port Said, across the Med. and up to Southampton with no further stop. I believe the ship's cruising speed was around 20 knots, propelled by two thumping 10-cylinder marine diesel engines. Junior officers were in cabins for four. This proved to be restrictive as most of us were carrying quantities of food and other non-uniform items which would otherwise be difficult to get back home, our hard luggage being in the hold. Our discomfort was increased when, within twenty-four hours of our departure, one of us dropped a hard object into the one basin in the cabin and knocked the bottom out of it. For the whole of the voyage we had to use the communal washroom which, I seem to remember, was fed from the sea-water system. Without the focal point of a unit such as the MONAB that we had had on the outward journey, daily life had no real object. I don't think we even had daily boat drill. Several of my friends were on board, Mac Rutherford and Johnnie Clemes being two, so there was common ground for conversation. We had expected the ship to be "dry" and, indeed, the bars were not open for alcoholic drinks. However, OC Troops allowed us to bring our own alcohol on board at Fremantle, a concession which was duly exercised. Many were the bottles of spirits and packs of beer which came aboard there. Having a day in Perth, we were also able to send home more parcels of food. It was not long before all the booze was consumed; no doubt many of the bottles ended up over the side of the ship. To keep us amused there were frequent lectures in the officers' lounge. Probably causing as much amusement were the efforts of certain officers to clinch relationships with members of the hospital and nursing staff of the Q.A. and other Wren officers from the staff in Sydney, who were now also on their way home. Bridge schools were soon formed and hours were spent at the bridge tables. Others preferred to find a sunny place up on the boat deck with a book. Food was good and plentiful. We had one or two civilian passengers amongst us, one being Norman von Nida, the Australian professional golfer, who was coming to England to take part in the annual golf competitions in Europe. Each day he would come up on deck with a box of golf balls, many of which ended up in the sea, and practise his swing. I was surprised that, as on the outward voyage, no effort was made to set up a visit from Neptune[9] as we "crossed the line". I had looked forward to this.

I had time to ruminate on my time in Australia. I was lucky to have had the experiences that I did. I was probably the only officer to have been on the four operational Air Stations in the country, and certainly the only junior officer to have been his own boss for so long. From my position with my unit MS1 I saw the Air Stores organisation as probably no other. In August 1945 my civilian contact in the Sydney office took me over to their depot in a new warehouse at Randwick. Even then, as the war was ending, the building was only half full. I appreciate that the problems were immense, not least the difficulties of storing spare parts for so many different types of aircraft. I have not seen a report of the Naval Air Stores system during the war; I would find it fascinating. The transition to war so soon after independence from the RAF must have been a tremendous shock, made somewhat easier by the adoption of the RAF stores system in its entirety. The contrast between the Naval Stores book and the Air Publications parts lists was a revelation.

Bearing in mind that the work of the MONAB organisation bore little relation to the original concept, the outcome in Australia and in the Admiralty Islands was remarkable. Admittedly we operated in a

9 Neptune of course is the mythical God of the sea, to whom mariners crossing the equator for the first time pay homage, usually by dressing up in silly clothes and getting tarred and feathered by those more seasoned fellow travellers.

particular circle centred on no.1 Aircraft Carrier Squadron. It wasn't until many years later that I learnt through such books as John Winton's *The Forgotten Fleet* how many warships there were in the British Pacific Fleet, protecting and supporting the carriers and the Fleet Train. There had been 155 sea-going warships, including the carrier squadron. The Fleet Train itself amassed 102 vessels of a multitude of capabilities. I take my hat off to all those who set up such a huge organisation. One should also understand that the East Indies Fleet, based in Colombo, was in existence at the same time and had been fully operational up to VJ-Day.

One of the many good things done by the Royal Navy to keep us all happy was the performance of the Fleet Mail Office. As soon as we were out there an air mail route was opened, primarily, I believe, operated by civil-marked Lancastrian aircraft which also carried a few passengers. The route was across Australia, the Indian Ocean to Ceylon, Bombay or Karachi, across the Middle East to Cairo, thence via Gibraltar to Hurn in Hampshire. By mid-1945 I could get a reply back from my parents in Edinburgh within a fortnight.

Regarding Australia itself, I was frequently struck by how different the countryside was from the towns and cities. But in spite of differences caused by the geography and the climate, the British influence was everywhere, and I felt very comfortable in my personal contacts. Australians were very conscious of the huge potential of their country, but aware of the handicap of such a small population - 7,500,000 in 1946 - for such a huge place. The country had suffered much during the great depression. The British Government agreed that the Sydney Harbour bridge should be built to bring both sides of the harbour together and offered money to pay for it. Although the steel components were all built by Dorman Long & Co. in Middlesbrough in the North-East of England, much economic activity was generated in New South Wales, particularly in the quarrying and dressing of suitable stone for the massive towers on which to mount the steelwork. The bridge was completed in 1932. There was an economic bonus for the North-East of England because work and jobs were created in the steelworks, which were also suffering from the depression. In 1946 Australians were already talking about an extensive scheme to divert water from one side of the Snowy Mountains of New South Wales to the other, both for a hydroelectric scheme and to add to the river flow on the west side of the Great Dividing Range into the Murray-Darling basin. Starting in 1949 it took twenty-five years to complete. They also realised that they should start to standardise the state railway gauges to enable inter-state trains to run without interruption. An example of the resultant disruption was in Albury, on the main line between Sydney and Melbourne. The two State Railway Companies had built the longest roofed passenger platform in Australia there. On one side was the NSW standard-gauge track, on the other the Victoria State Railway's 5 ft. 3 in. gauge. Trains had matching carriage and seat numbers and were pulled up with these numbers opposite each other; passengers just walked across the platform to take up their new seats. With the conversion of the line from Albury to Melbourne to standard gauge the procedure was no longer necessary. The internal railways of Queensland, such as Brisbane to Maryborough, are built to 3 ft. 6 ins. gauge, but the NSW State standard gauge network included a line from Sydney to Brisbane. Thus our Naval personnel travelling south from Maryborough could change gauges at Brisbane.

The hospitality the Australians offered to the Royal Navy was extraordinarily generous, both in their families and in their social clubs – for instance, Naval officers were welcomed to the clubhouse of the Royal Sydney Yacht Club in Macquarie Street for a drink or a meal. Service people were also welcomed in the numerous Returned Servicemen's League clubs. Such hospitality was evident in many cities and towns. It could be embarrassing, so effusive was the welcome, and so immediate. But we loved it! We were introduced to surfing which at that time was known as "body surfing"; the surfing board had not yet been invented.

The one and only dark side of Australian civilian life which we encountered was the behaviour of the dock workers in Sydney. They were notorious for their frequent "withdrawal of labour" and this they did on the arrival of the BPF in February 1945. Many members of the crews of our warships had to carry out dockside tasks while the "dockies" were on strike. They thought that the current Labour government would be sympathetic but I don't think this was the case. I had heard that earlier in the war the coalminers in the Wollongong, NSW, area had challenged the government by going on strike; the riposte had been to draft all the miners into the Army and order them to go down the pits, on threat of court martial.

As we thumped our way across the world, driven by our huge diesel engines, we had time to think about the future. I had enjoyed my four and a half years in uniform but I did not think I would fit into peacetime service. If I was to enter the aircraft industry it was important to match my ambition with my capabilities and my experience. One way and another I had already been connected with aeroplanes for some twelve years and my enthusiasm had not waned. My wish to turn my back on university education was encouraged when, a few months before, my father had written to say that the entry list for engineering in Edinburgh University for the 1946 academic year was already full. I was blowed if I would wait until 1947 to start another three years of theoretical engineering. During my war service I had come across a number of air engineer officers who had been through the De Havilland Aeronautical Technical School at Hatfield and Stag Lane. All of them were very nice people and competent at their jobs. They had an aura about them. By the time I was back home in April I had decided that I would approach DH's and inquire about the content of their courses. I was not the only one to be thinking about the future. Johnnie Clemes, who, it transpired, was a very good mathematician (and a very good bridge player), decided to abandon aeronautical engineering and become a chartered accountant. This led him to the top of his profession and he became finance director of a large public company. We all felt that our wartime service had put us back in developing our lifetime careers and that it would be hard to catch up.

After our day in Perth, WA, we saw no land until the ship anchored off Aden for an exchange of mail and the embarkation of several British civilians, probably important Colonial Office officers. In the Suez Canal we passed an eastbound troopship, *Alcantara* I believe, and we spent another day at anchor in Port Said. The ship was surrounded by "bum boats" selling all sorts of dubious products which might improve the enjoyment of our voyage. One of these was Cyprus gin, suspected of having questionable additives. What we did for money, I cannot remember, but I expect our Australian currency had been withdrawn before we left the continent, in exchange for Sterling. Mail came aboard at Port Said and I was surprised to be summoned to the purser's office to collect a letter. It was from my parents in Edinburgh and was the only letter for a Service passenger. Drawing on his, rather dated, knowledge of Service procedure, my father had sent just one more letter headed by my rank and name "c/o GPO London". The Fleet Mail Office evidently already knew that I was on my way home in *Stirling Castle* so to Port Said the letter went. It was my last experience of the extraordinary efficiency of the Fleet Mail Office.

It was in the early morning of either 14th or 15th April 1946 that we slowly edged into the Solent under a light mist, fog-horn blowing – its first blast startled all of us who had gone up on deck early. We were anxious to see our home country for the first time after seventeen months. There was no crowd to welcome us as we came alongside the famous ocean quay at Southampton. It has been commented upon by historians that there was never a welcome for any ships of the BPF or for the troopships which brought thousands home, including released POWs. The complications of the end of the war in the Far East and the recovery of British Colonies and interests meant that repatriation took many months. Ships were going to and fro for a considerable time after the shooting had stopped, and the implementation of the Age & Service formula had resulted in a steady release of men and women for probably as long as eighteen months, perhaps more. Travelling slowly in the train through Southampton on its way to London I was delighted to see numerous flowering trees in the back gardens of the houses, harbingers of Spring.

So, after drawing my civvy outfit from the huge "demob" depot at Wembley Stadium, I finally reached home in Edinburgh about the 16th April. My active service ended on 30th June 1946.

Now – that job! I did get a place at the DH Aeronautical Technical School at Hatfield as an ex-Service student to study Production Engineering on a three-year course. This also included an indentured apprenticeship. A grateful country paid my fees and a small living allowance. What a difference from university!

De Havilland Aeronautical Technical School

Using my gratuity – a "thank you" for the service I had given my country – I bought a new AJS motorcycle when I was at home in Edinburgh and rode back and forth to the south of England several times during and after my terminal leave. I wished to secure a place as a premium apprentice in the De Havilland Aeronautical Technical School at Hatfield.

The course for which I was accepted was in Production Engineering. Schools such as these were to all intents and purposes technical universities and were set up in a very professional manner, coupling practical workshop training with evening and day release classes. The syllabus for these covered subjects in the examinations for membership of a professional body, in my case the Institution of Production Engineers. Others were the Royal Aeronautical Society and the Society of Licensed Aircraft Engineers. In these modern times such membership entitles the holder to the grand title of Chartered Engineer, equivalent to university degree status. Seventy years ago the majority of directors and managers in the engineering industry were educated in this manner and jobs were open to them all over the world.

Trade apprentices at De Havilland were enrolled in the year they completed their primary school education and were indentured in the traditional way, a fee being paid by the parents. At that time there was no nationally funded secondary education, only grammar and public schools, for which fees were paid. It was from the latter two categories that the DH premium apprentices (known colloquially as 'students') were drawn. All were expected to have passed their School Certificate so would be between 17 and 19. There was an entry examination and fees were paid for the three-year course at the Tech. School. As well as undertaking a course for a professional qualification students were indentured as apprentices. There was a scheme which enabled successful trade apprentices to be upgraded to 'student' after their first year and proceed to study for professional exams.

The Tech. School Principal was S/Ldr. Reeve who had been on the DH company staff in the 1930s as the Principal and Chief Flying Instructor of the DH pilot training school. This was affiliated to the RAF as No. 1 Elementary & Reserve Flying Training School. In charge of day-to-day administration of the boys' activities and factory placements was Mr. Blackburn.

Both categories of apprentice spent the first year in the rather grand setting of Salisbury Hall, which was a very historic 17th century mansion which had been taken over by the DH Company as the temporary home of the drawing and production offices in 1941. It was some eight miles to the south-west of the factory. Separately an aircraft hangar had been built which was used for the assembly of the first two Mosquito prototypes.

This building was subsequently used as the Tech. School workshop which was set up with all the appropriate equipment; lathes, milling machines, pillar drills, woodworking machinery, benches for precision hand tool work, sheet metalwork and panel beating. The specification and use of hand tools was a classroom subject. Every student made his own wooden toolbox, and the school supplied a selection of tools to fill it. Several practical exercises took the form of an item which could be used on the bench or in general engineering situations, e.g. an oil can, a funnel, precision blocks and clamps for milling work.

It was refreshing to be in the company of young men who, only a few months before, had been schoolboys. I was lucky to have temporary accommodation in the students' hostel, Sherrards House, at Digswell, two or three miles north-east of the factory, beside the A1 main road. This was a substantial Victorian mansion which was taken over by the De Havilland company during the war. It was very convenient for the factory, but it was 8 – 10 miles from Salisbury Hall. The students were all mad on motorbikes and cars – and aeroplanes, and there was a noisy mass exodus after breakfast. Inevitably students made the run to Salisbury Hall a competitive event and during my three-month stay there the record was held by a young man called Mackintosh who had a Triumph Tiger 80 twin (this may not be the exact description, but it was a Triumph twin.) Sherrards was full of laughter and high spirits – the dormitory arrangements were familiar to those who had been to boarding school – and the late evenings were filled with the nightly broadcast of a big band programme from the US Armed Forces' in Europe radio station.

The worst of the pranks committed, which I never saw, was to dismantle one of the motorbikes and spread the parts around the dormitories; the wretched owner had to hunt for the parts and re-assemble his bike before departing for Salisbury Hall the following morning. My status as an ex-serviceman entitled me to a small single room off a passage which kept me away from the hubbub elsewhere in the house. A bonus for me was that I had met the husband of the lady who ran the hostel in the officers' mess at Henlow in 1942 and we recognised each other. He had been a catering officer in the RAF and was the manager of the Company's catering department.

There was great enthusiasm at the start of Year 2 as we were all moved to the factory at Hatfield. De Havilland had been in the forefront of the transition to jet propulsion. The engine company at Stag Lane had engaged Frank Halford to design a jet engine to Frank Whittle's concepts and this became the Goblin. Surprisingly engines were ready for flight before the Rolls-Royce production version of the Whittle engine and powered the Gloster Meteor prototype on its first flight. The Hatfield factory was still building propeller-driven military aircraft such as the last of the Mosquitoes, the Hornet twin-engined fighter and the DH Canada's military trainer, the Chipmunk., but the new order in military aviation was to be seen in the Vampire single-seat jet fighter which had entered squadron service in the RAF in 1946.

It was also building the first small twin-engined civil aircraft known as the Dove. There was plenty to interest the students; in fact, with technical schoolboys all over the factory there was a very active grapevine to keep everybody interested. Going back a year, however, my arrival had been at the time of the catastrophe which befell Geoffrey de Havilland in September 1946. The tail-less DH 108 experimental aircraft had disintegrated in a transonic dive over the Thames estuary. Geoffrey, the Company's chief test pilot, was the second of Sir Geoffrey's sons to die whilst testing one of his company's aircraft. Also in the background, in the experimental shop, was the construction of the first Comet jet airliner, eventually to fly in 1949. For all this activity the Hatfield factory still had jobs for thousands of men and women.

There was an active Sports and Social Club at Hatfield with its own premises beside the football and rugger pitches at the east end of the factory grounds. I joined the rugger club and enjoyed a game every Saturday over the winter months. The standard of play was not as high as the main London clubs but was equal to their second teams and a number of Old Boys' teams from the large number of day schools around London. There were also other factory club teams and local Hertfordshire town clubs. Hertfordshire RFC was a second-line county team which took part in an annual county competition. I was lucky to be picked for one of their matches.

Partly, I suppose, for flexibility in the use of the factory floor, the main factory building had virtually no partitions, so manual sub-assembly was in an area within sight of the aircraft assembly line. To start the production process the main components of the aircraft such as the fuselage and the wings came together on a wheeled framework until, along the line, the undercarriage was fitted and the emerging aircraft could be moved on its own wheels. At the final stage the complete aircraft went straight forward into the paint shop before being drawn outside by tractor for its first engine run. There was a mixed line of types going through at once so we could hear a variety of engines being run up, sounds which stimulated our interest. It was exciting to see things actually happening day by day.

Huts along the front of the factory had been allocated to the school as lecture rooms. Being an independent organisation, the school was responsible for engaging the teaching staff, all of whom came from within the factory. One could say that all the subjects together would provide the basis for setting up and running a factory: finance, book-keeping, balance sheets, factory layout, manufacturing methods, performance of materials, personnel management, job specifications, management structures and so on. I found it all fascinating and a complete contrast to the syllabus for mechanical engineering at university. I think there may have been a mathematics class which was well within my capabilities. Exams loomed up in June or July 1949. A choice of paper was allowed and I chose metallurgy, a subject which I had found difficult at Edinburgh and at Henlow. However, I felt I had got on top of the subject and was confident that I could please the examiner. I passed well!

Success in the examination as a whole gave me the status of Graduate of the Institution of Production Engineers. I had also fulfilled my indentures and received a handsome certificate to prove it. I wanted to remain with the company to get experience in one of the departments within the production process, and Mr. Blackburn found me a position in an embryo Development department under Mr. Long. This was new and had been set up to examine and test materials for the new Comet. Conditions of extreme temperature and pressure contrasts were being encountered at the high altitudes at which the aircraft flew so we had to test the materials, using low temperature refrigeration. This work followed one of the most thrilling experiences of my life.

It was the rule that the factory production line shut down during the Company's annual holiday, a break of a fortnight. Students arranged with the head of the department they were working in to stay at work and carry out non-standard jobs. This was planned to be the time when the Comet flew for the first time, a fact known to the student body who, to a man, arranged their holidays for another time. It was a lovely day on 27th July 1949 as factory workers and students took their lunchtime sandwiches outside – just to enjoy the sunshine, of course. The wait was agonising as John Cunningham and his crew of observers ran the engines as they went through all their ground checks. The complete board of directors came out and stood not far away from me, fidgeting with anxiety. Then an extraordinary thing happened. I could feel what I can only describe as an electric charge welling up in the spectators, almost compelling the aeroplane to take off safely. Then, as Cunningham opened up the engines, and the machine started to move, this feeling became almost overpowering. But, miraculously, as soon as it left the ground, it vanished! The Comet had flown! In my life this effect only happened to me once again, at a motor race at Goodwood in 1954. My old friend Ron Flockhart was driving the V-16 BRM in a race in which the main opposition was the V-12 Thinwall Ferrari driven by Peter Collins. For two or three laps the cars were neck and neck and I experienced the same electric charge as at Hatfield. Collins eventually drew away by a few seconds and, as he did, so the charge dissipated.

Towards the end of 1949 I assessed my future. I wanted to stay in the aircraft industry but the salary levels were poor. De Havilland had a good production programme, with versions of the Dove light transport, the final propeller-driven fighters, the Hornet and Sea Hornet, Vampire and Venom jet fighters for the RAF, the Royal Navy and a number of foreign air forces. The engine company was building Goblin and Ghost jet engines for the company's aeroplanes. However, the country was effectively broke and the future of our armed forces looked grim.

Orders for four large Malta-class fleet carriers were cancelled in October 1945 before building started. Other classes of fleet and light carrier were ordered and many ships were laid down, but very few were finally commissioned in the Royal Navy. I had thought of putting myself forward for some form of permanent commission in the Navy now that I had a worthwhile qualification, but I realised that I might only be offered a short service commission which would see me back on the jobs market around the age of thirty.

Regretfully, early in 1950 I decided to leave the aircraft industry; I ended up as a management trainee in the corrugated cardboard box industry. However, that year I was enticed back into the RNVR and was assigned to the London Division reserve squadron, No. 1832, based at RNAS Culham, south of Oxford. After all I had retained a slender hold on aeroplanes and the Royal Navy.

Postscript

The object of this memoir was in the first instance to recall my wartime service. However, a long life was to follow which took unexpected turns. In terms of the "day job" the major part has been connected with the motor trade, first in a retail car dealership in Ascot, Berkshire, then in wholesale parts distribution from our family home in Pembrokeshire.

Although my RNVR service was to end in 1952, thereafter I maintained a strong relationship with naval aviation through active membership of the Fleet Air Arm Officers' Association (FAAOA) and the Society of Friends of the Fleet Air Arm Museum at RNAS Yeovilton. I joined the latter in 1980, shortly after it was formed, and was given the membership number 148. At that time I donated a number of books and documents to the new library in the museum. I also joined the historical organisation Air Britain and struck up a friendship with Ray Sturtivant, an experienced and dedicated historian and writer of naval aviation history. My contribution to his work was particularly based on my service in Australia, in MONAB1 and 706 Squadron, and subsequent correspondence with pilots and observers who had served in squadrons in the carriers of the BPF. The information I obtained from them has proved to be invaluable.

Combining reading in my gradually extending collection of books and other archive material enabled me to write short articles on naval aviation matters, British and American, specialising in the Pacific war. I lectured to the Friends of the FAA Museum and three of my articles were published in their periodical magazine.

Regarding the FAAOA, I was particularly proud to have acted as Honorary Secretary of the South Midlands Group of the Association for almost twenty years from 1995. I took over from a famous and much-loved FAA character, Major Teddy "Cheese" Cheesman DSO MBE DSC, Royal Marines (Retd.) who had started the group in 1977. The creation of this system of groups went hand-in-hand with the start of a programme of monthly luncheon meetings in a pub, known as First of the Month Meeting or FOTMM. These were great occasions for recalling old times and spinning unlikely tales. I must have supervised over 200 meetings of our group, some of which were special commemorations and included services in the local church. That period was perhaps the zenith of our Association. Ordinary lunches would attract between twenty and thirty people – wives included, and on the big occasions we would see as many as ninety. It has been a melancholy experience to see the numbers drop to below ten as age has taken its toll.

The caption to the group photograph of No. 1 course of the Edinburgh University Air Squadron on **page 6** records the presence of Ron Flockhart. We had both started the mechanical engineering course at the university in 1940 and immediately became great friends. This was largely due to our common interests in motoring sport – and he had a motorbike which on numerous occasions provided me with pillion transport. In 1950, after my father had died and my mother had moved back to her native county of Kent, I was once more in Edinburgh, the Flockhart family's home for generations. I lived in "digs" but spent a large part of my spare time in Ron's home, talking motor cars and bikes. We also talked about going into the motor trade, a plunge which was eventually taken nine years later. He had committed himself to motor racing as an amateur, moving into the slender ranks of the professionals in 1954 with BRM. Ron was one of the first to be appointed as a team test and race driver. Two years later he joined the Ecurie Ecosse sports car racing team, a small organisation based in Edinburgh, the home city of the team proprietor David Murray. It was with this team that Ron won the Le Mans 24-hour race in 1956 and 1957. He had little success driving BRM cars at home but was sought after by other owners to drive their cars, with some success.

Ron and I eventually went into business in the retail motor trade, buying a small garage and retail dealership for Standard-Triumph in Ascot High Street, Berkshire, at the beginning of 1959. This was really a holding operation for Ron whose activities in motor racing precluded active participation in this business. By this time, he was an experienced pilot of light aircraft and had his own Auster. He had been to Australia and New Zealand for the local summer series of motor races. A BRM was shipped out specially for him to drive and with this car and others he had some success. While there he put into effect a plan we had discussed at home over an ordinary atlas – to buy an ex-RAAF Mustang fighter and fly it back to England in record time. The first effort, in 1961, ended at Athens with an unserviceable Merlin engine, the second, in April 1962, ended in personal disaster when, in poor weather, Ron crashed into the bush on the Dandenong mountains near Melbourne, losing his life.

This unexpected disaster was to influence my subsequent decision to sell the Ascot property, by then worth a considerable sum of money, and to move to Pembrokeshire with my wife and two children. There I operated a very personal wholesale motor parts company for over twenty years – and that is a story in itself. The final chapter, still unfinished, was removal back to England in 1994 to settle in the small town of Bromyard, Herefordshire.

A long life and a vivid memory served to bring publicity in 2019 when reminiscences of my association with Ron Flockhart, Ecurie Ecosse and Le Mans were recorded in a video by Richard Suckling, shown on YouTube, and an interview with Gordon Cruickshank for the magazine Motorsport. Fame after sixty-three years!

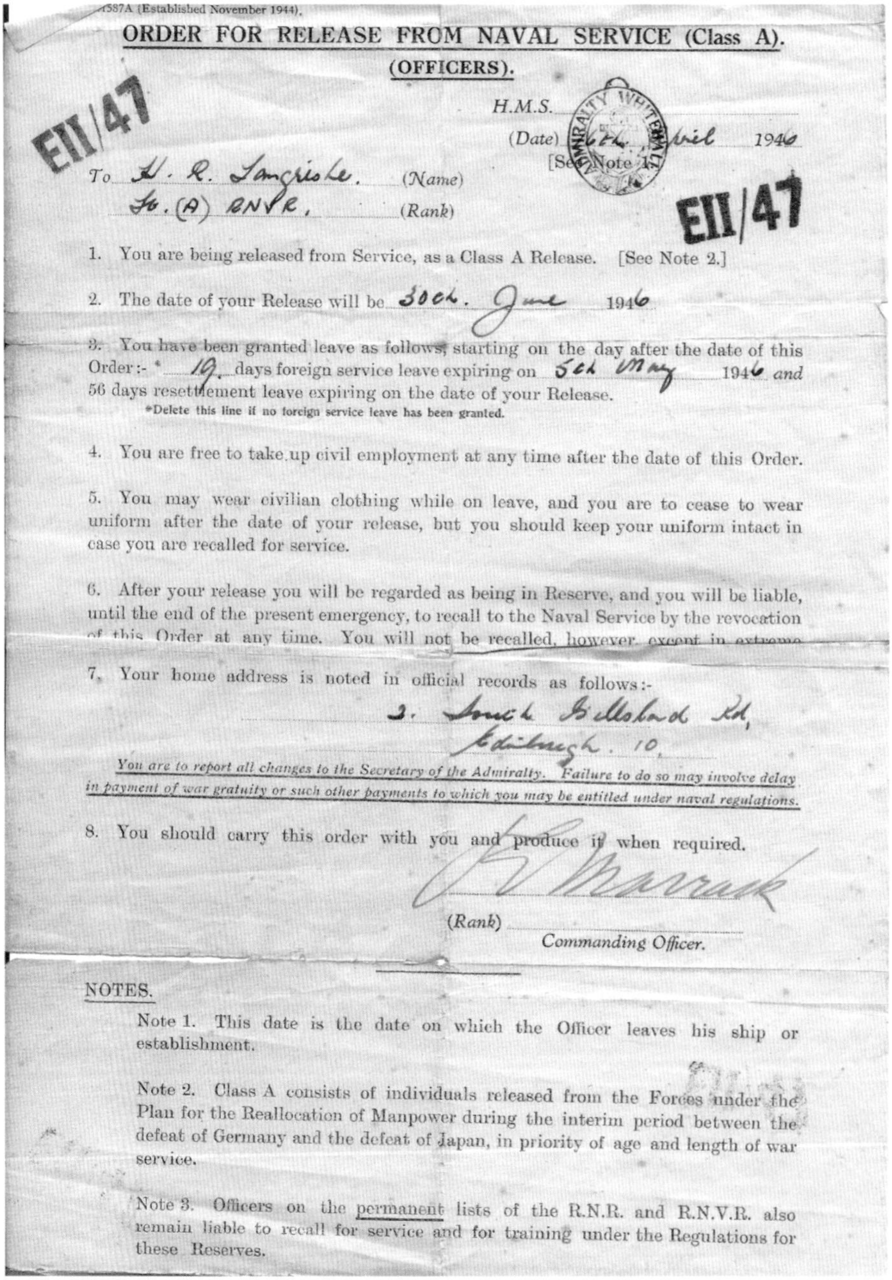

S.1587A (Established November 1944).

ORDER FOR RELEASE FROM NAVAL SERVICE (Class A).
(OFFICERS).

E11/47

H.M.S. ....
(Date) 6th April 1946
[See Note 1.]

To H. R. Langrishe. (Name)
Lt. (A) RNVR. (Rank)

E11/47

1. You are being released from Service, as a Class A Release. [See Note 2.]

2. The date of your Release will be 30th. June 1946

3. You have been granted leave as follows, starting on the day after the date of this Order:- * 19. days foreign service leave expiring on 5th May 1946 and 56 days resettlement leave expiring on the date of your Release.
*Delete this line if no foreign service leave has been granted.

4. You are free to take up civil employment at any time after the date of this Order.

5. You may wear civilian clothing while on leave, and you are to cease to wear uniform after the date of your release, but you should keep your uniform intact in case you are recalled for service.

6. After your release you will be regarded as being in Reserve, and you will be liable, until the end of the present emergency, to recall to the Naval Service by the revocation of this Order at any time. You will not be recalled, however, except in extreme

7. Your home address is noted in official records as follows:-
3. South Gillsland Rd,
Edinburgh. 10.

You are to report all changes to the Secretary of the Admiralty. Failure to do so may involve delay in payment of war gratuity or such other payments to which you may be entitled under naval regulations.

8. You should carry this order with you and produce it when required.

[illegible signature]
(Rank) ....
Commanding Officer.

NOTES.

Note 1. This date is the date on which the Officer leaves his ship or establishment.

Note 2. Class A consists of individuals released from the Forces under the Plan for the Reallocation of Manpower during the interim period between the defeat of Germany and the defeat of Japan, in priority of age and length of war service.

Note 3. Officers on the permanent lists of the R.N.R. and R.N.V.R. also remain liable to recall for service and for training under the Regulations for these Reserves.

Although issued under the terms of the Age and Service release regulations there is no obvious place for the entry of the appropriate category number; there must be some significance of the number "E11/47".

APPENDICES

Appendix A: Edinburgh University Air Squadron, 1941–1942

The website for the Edinburgh University Air Squadron features the Squadron over the years 1946-1969. There is reference to its formation during the war, but only to say "- initially formed in 1941 at RAF Turnhouse but after merging with Glasgow and Aberdeen was disbanded early in 1946." The Air Britain book "Royal Air Force Flying Training and Support Units" records "Formed 1.41 and attached to Turnhouse with 1 Tiger Moth from 4.41".

Neither of these records is accurate or sufficiently informative. Nationally the Air Squadrons were formed at a number of provincial universities early in 1941, not to provide flying training but to give undergraduates who might wish to curtail their university course an opportunity to enter the RAF or the FAA to serve their country in the war. The course which was adopted had the same content as the Initial Training Wings of the RAF and of the RN establishment *HMS St. Vincent* at Gosport. This stage was, and is, the ground school which precedes initial flying training. During the early months no aircraft was issued to the squadrons; in the case of Edinburgh, the Miles Magister of the RAF Turnhouse Station Flight was lent to the squadron for air experience flights, piloted by the squadron CO. The university squadron had no technical aircraftmen to service an aircraft.

Squadron manning was simple yet complicated. The only serving members were the CO, S/Ldr H.M.Pinfold RAF, a career officer who had participated in the Battle of Britain and was to return to the front line and be killed in action, adjutant F/L Mackay RAFVR, Flight Sergeant Downie the Chief instructor and three Corporal instructors. The student members of the squadron were not enrolled in the RAFVR but were Air Cadets; the uniforms had Air Training Corps buttons but there were no official numbers and no status in the RAF. Entry to the squadron was by interview by the incoming CO but there was no medical examination. Service discipline was followed but in practice could not be enforced because the squadron members were civilians. So far as I am aware there was no ATC Squadron number and there was never any contact with the local squadrons of the ATC in Edinburgh. Turnhouse had no part in the day-to-day work of the squadron and had no supervision of the squadron's activities.

In the early months of 1941, the squadron came into existence in premises allotted by the university, the gymnasium of the Physical Education department. This was a good choice. The Head of Department was Col. R. Campbell, a retired regular soldier who was dedicated to his country as well as to his students. The gym was in effect new. It was a two-storey facility built into an old stone building and had only recently been completed. The upper floor was just a large open space. The lower floor, of a size large enough to accommodate a badminton or a basketball court, was fully equipped with the standard gym apparatus. The building was beside an ancient city street known as Pleasance, a short distance from Holyrood Park and Arthur's Seat. The Squadron enjoyed PT under the leadership of the two permanent instructors. To render it suitable the upper floor of the gym was divided into three or four classrooms by building temporary partitions on the polished hardwood floor. Offices were also provided for the CO, the adjutant and the NCOs.

The first parade was held in one of the university lecture theatres close to the gym, when the C.O. addressed the cadets. He told us of his background and his experiences in the front line the previous year, and what the aims and objects of the squadron were. Within a short time, the dedicated premises were opened and training began; there was great enthusiasm. The subjects taught were:

- Aircraft Recognition
- Air Navigation by Dead Reckoning
- Anti-gas
- Armament (with practical use of the Vickers 'K' machine-gun on the Arthur's Seat open range)
- Hygiene
- Mathematics
- Service Law
- Signals (Wireless Telegraphy)
- Parade ground drill was practised in the lower gym under the control of "Flight"

During university term-time classes were held in the evenings but I think we turned to full-day classes during the university Easter vacation. Examinations were held at the end of the course in July and marks were communicated to the cadets by individual reports. Without a parade ground, there was no formal passing-out parade. I can remember no occasion of an inspection visit by a senior RAF officer during No. 1 Course.

Where's Hugh?
On floor in middle: S/Ldr. H.M.Pinfold RAF, Squadron Commander.
3rd row back: HRL 7th from left; Ron Flockhart 6th from right:
2nd row back: J.M.Bruce 4th from left.

Immediately after the end of the course, during July 1941, all cadets went through a Flying Medical Exam. Not all cadets wished to join up straightaway and, in any case, such things did not happen overnight. My attention to my studies had been seriously affected by my interest in "the squadron" and I did badly in my university exams. My father persuaded me to re-sit in September, but my heart was not in it and the results were no better. I committed myself to the squadron and in the rank of Cadet Corporal, granted by the CO, I became a member of the teaching staff, taking the Aircraft Recognition classes. I produced much more information for the classes than the RAF could provide.

Following my call-up in March 1942, I visited the squadron four times during leave to my home in Edinburgh, on 3rd September 1942, 2nd November 1942, 19th April 1943 and 21st June 1943. At the first and second visits the staff remained the same but by the third the CO had been posted as had F/Sgt Downie. The corporals were also expecting to be posted soon. There was a rumour that the squadron premises were to be taken over by the Naval Division and the squadron moved to other premises in George Square.

By the visit on 21st June the squadron was still in the Pleasance premises and F/L Mackay was still the adjutant. However I did not record the name of the new CO and that was my last contact with the squadron.

Appendix B: RAF School of Aeronautical Engineering, No 16 (War) Course

Lectures and demonstrations; examinations. 9th July 1942 to 29th December 1942

[Where not indicated, and title is given item is a lecture]

9th July 1942	Introduction to Course: CO and Adjutant
10th July	Film: "Principles of Flight"
11th July	"Electricity".
13th July	Individual interviews with CO
	"Aerodynamics" – the Professor.
	Drawing Office: 2 hrs. sketching of Napier Lion piston.
	Film: Aircraft Recognition – Spitfire I.
14th July	"Internal Combustion Engines".
15th July	"RAF Organisation & Administration". RAF officers only.
16th July	"Metallurgy".
17th July	"Aerodynamics".
17th July	Film: Aircraft Recognition – Messerschmitt Me109B.
18th July	"Metallurgy".
20th July	Drawing Office.
	Course student lecture: H.R.L. on "Aircraft Recognition"
21st July	"Heat Engines".
23rd July	"Electricity".
24th July	"Aerodynamics".

1st August Discussion: "Mass Production".

(On or about 2nd September No. 14 War Course passed out, replaced by No. 17.)

At that point I stopped recording the lecture titles. As examinations were not until October, and again at the end of the course, it is probable that a timetable of lectures in the subjects being taught was continued throughout the course. Many afternoons were spent visiting the workshops of 13 MU with the guidance of our workshop instructors. These visits are recorded later.

Interim examinations took place as follows

5th October Metallurgy – 51%.
7th October Aerodynamics – 75%.
8th October Heat Engines – 68%
9th October Electricity – no marks recorded.

Lectures

20th October "Planned Maintenance".
26th October "Faulty Maintenance".
5th November Lectures to be held at 08.30. No other details recorded.
13th November "Plastics" (outside lecturer).
20th November "Sparking Plugs" – by a Lodge Plugs representative.
27th November "Alloy Metals" – by a representative of High Duty Alloys.
4th December "Fighter Design" – by Frank Lloyd of Hawker Aircraft Co.
23rd December "Heat Engines" ("Last lecture", so there must have been more.)

Final examinations took place as follows:

10th December O and A – 64%
11th December Metallurgy – 50%
14th December Electricity – 65%
15th December Structures – 63%
16th December Heat Engines – no marks recorded.

16 War Course; Practical Workshops

Time spent in each session varied but normally was most of the morning, Monday to Saturday. Sometimes the session was interrupted by a visit to the appropriate shop in 13MU, led by the instructor. Leave periods fill gaps in dates.

9th to 22nd July. Panel beating and repair (Mr. White): Shaping and wiring aluminium cup, repairing petrol tank, repair skin of Walrus hull, making copper funnel.

23rd July to 5th Aug. Machine shop: Machining gear from cast iron blank, turning to size, cutting teeth and thread cutting.

6th to 12th Aug. Welding (Mr. George Fuller): Gas and electric arc welding of steel and aluminium, fillet and flange welding. Demonstration of welding aluminium castings and magnesium sheet.

13th to 15th Aug.	Forge (blacksmithing): Hammering out blanks for scriber and spanner, scarf welding.
17th to 19th Aug.	Foundry (Mr. "Blow-hole Bill" Quinton): Making sand mould for a ring, a bearing and then for a model Spitfire, melting aluminium and casting model.
20th to 26th Aug.	Fitting shop (Mr. Gascoigne). Precision use of files and hand drills, making paper punch.
27th Aug. to 2nd Sept.	Lathe shop (Mr. Goater): Range of lathe work, screw-thread cutting.
7th Sept. to 3rd Oct.	ARS (Mr. Borritt & others): Fabric repairs (four sessions), wire splicing, introduction to aircraft instruments (Mr. Elms), incl. visit to 13 MU instrument repair shop, duralumin tube repair and extrusion repair on Hawker Fury fuselage structure, flush sheet repair on Blenheim tailplane; to 13 MU to examine a Typhoon fighter in pieces (Flt Sgt. Armstrong). Strip and rebuild Rotol c/s prop and DH bracket-type prop, theory of electric props; to 13 MU propeller shop and Perspex repairs, studied Exactor hydraulic control system, use of wind tunnel with brass aerofoil, calculations of drag and lift, hydraulic systems and components (Wellesley hydraulics), pneumatic systems, flotation gear and rescue apparatus, tour of 13 MU ARS – Hudson and Hurricane repair lines.
5th to 31st Oct.	ERS (Mr. Bosworth): (First five days interrupted by exams.) Liquid-cooled section: RR Merlin: dismantled, examined and discussed component parts – oil pump, automatic boost control, hand and electric starting gear train, gear case and supercharger gears; valve and magneto timing, Stromberg injection carburettor. Air-cooled engine section: film show on magnetos fitted to Bristol engines and on removing cylinders from sleeve-valve engines. Strip and rebuild Pegasus XVIII and the component parts - reduction gear, two-speed supercharger and clutches, valve gear, rear cover and accessories, crankshaft, talk on Claudel Hobson carburettors; timed magnetos. Dismantled G.E.C. turbo-supercharger. Tour of 13 MU ERS included the MT workshops, overhaul of auxiliary pumps etc. Ran Gipsy II on test bed. Crated Cyclone engine for transit to Lee-on-Solent.
5th to 18th Nov	Electrical lab. and workshop (Mr. Eaton). Spark plug cleaning and setting; automatic control systems; magnetos; engine starters, generators and regulators, carbon piles; battery chargers; automatic bombsights; Sperry gyro-pilot; MT electrical systems; aerodrome neon beacons and airfield floodlights, electro-plating and anodising.
19th Nov. to 2nd Dec.	"New" hangar (in eastern corner of airfield). (Mr. Forder). Fairey Battle and RR Merlin engine: removal and installation of engine, ground running and adjustment of carb. Blenheim; "dummy" inspection routine, starting and running-up engines.
3rd to 9th Dec.	Heat engine laboratory. Use of the Ricardo single-cylinder test engine for checking and comparing fuel octane numbers and for fuel consumption comparisons; supercharger testing; cathode ray oscillographic indicators on

	Ricardo test engine; visited test bed development flight, silent test house, Pegasus test before "duration run".
10th to 16th Dec.	Woodworking shop (Mr. Hornibrook). (This period interrupted by exams.) Use of hand tools and lathe.
17th to 24th Dec.	Heat treatment laboratory. Heating and cooling curves for carbon steel; polishing, etching and examining carbon steel micro-structure after heat treatment to various temperatures; mechanical testing of carbon steel specimen after heat treatment; examination of macro-structures, showing grain of metal after forging.
28th Dec.	Guided tour of GES workshops: machine shops, welding, coppersmiths, heat treatment, electro-plating. Guiding by Flight Sergeant and Sergeant, both very good
29th Dec. (Last day of Course.)	Guided tour of turret repair shop: Frazer-Nash hydraulic turrets; Consolidated and Martin electro-hydraulic turrets; servo ammunition feed systems. Link Trainer overhaul and testing – course students allowed half-an-hour's "flight".

At various times on the course we were allowed to go and examine aircraft on the tarmac; Mustang, Typhoon, Albemarle, Halifax, Stirling etc.

Appendix C: RAF School of Aeronautical Engineering, Air Electrical Course

The less said about the course, the better. For the first fortnight (22nd April to 6th May) there was no set programme, and we were left on our own for long periods most days. There was no regular workshop schedule as there had been on the engineering course, and no lecture programme on theory until 7th May. On 6th May (to 23rd May) we were allocated to the hangar to work on electrical systems in actual aircraft. The most impressive was a Short Stirling heavy bomber which was the first aircraft to fit electrical systems where most other aircraft had hydraulics. We also created an electrical system with cables, switches etc on a large board.

On 7th May there began a series of lectures by Flt Lt Agger for which we joined the new junior electrical course.

24th to 29th May. Link Trainer room (Mr. Bruin).

31st May to 10th June. Instruments (Sgt. Ashton).

11th to 12th June. Automatic controls and gyro instruments (Sgt. Madden). "Most instructive of the course; only five of us" so "more attention from the instructor".

17th June. Departed RAF Henlow for the last time.

Appendix D: Personnel of MS 1 (Avenger), MONAB1, February 1945

Lt (A) (A/E) Hugh R.Langrishe RNVR,

AA (A/E) Humfreys

LAF (A) Brown
LAF (E) Ellison
LAF (L) McGowan

AA (L/O) Bailey

PO RM Richardson) Posted
PO RM Morris (AW)) out
PO RM Young (AW)
PO RM Atwill (AR)
LRM Fielding (AR)) promoted to PO RM and posted out
LRM Brown (AR)) do.

These names are from my notebook, but it is not clear whether they are the original men, or did I change the notebook as new men came in? Replacement postings maintained the strength of the unit at 8 ratings.

Acknowledgements

From an early age I collected things: cigarette cards, albums of scraps from motor and aeroplane magazines, aeroplane photographs, lead soldiers and early 1:72 scale aeroplane models. Inevitably this led to writing things down into notebooks – aeroplane details particularly. Over the years this habit spread into the recording of incidents in my own life. Matters became rather more serious when war broke out; a hobby naturally became an important resource for aircraft recognition. I found myself something of an expert when I joined the Edinburgh University Air Squadron in 1941 and, still a cadet, I was asked to lecture on the subject.

I decided that I should record the details of my training as an Air Engineer Officer in the Fleet Air Arm and I persevered with this record for some eighteen months, until I was established as a junior AEO at RNAS Fearn in the north of Scotland. At the time I realised that even three years after the recovery from the RAF of the Fleet Air Arm the Royal Navy had a serious problem in managing its air force, and aircraft maintenance in particular. Routines had been transferred wholesale from the RAF, whether suitable or not, and there was a significant shortage of experienced senior ranks to take charge of workshops or just to run the administration. The management of battleships hung over everything.

Aviation history has always interested me and I have maintained contact with the Fleet Air Arm museum from its early days, contributing books, papers and essays on various matters. I became a member of the Society of Friends in 1960, a year after its foundation – my membership number is 148. On my return from Australia in 1946 I wrote several essays about Fleet Air Arm activities in the South Pacific area during the last nine months of the war and I consider these unique in telling what actually happened, regardless of what London said.

I thought it would be a good idea to gather together all my wartime notes, without any thought of publication, and to create a memoir. Progress was delayed twenty years ago by my participation in another history project, the creation and fitting out of new premises for the Bromyard and District Local History Society. However, in the meantime I met two gentlemen harbouring the greatest enthusiasm for my personal project, each approaching from different directions. The first was Nick Blackman who was about to complete a full career as an Air Engineer Officer in the rank of Captain RN. He has for many years looked closely at the history of his branch and was enthused when I unexpectedly emerged from the woodwork. Nick was the influence which spurred me on. I asked him to look critically at my work and to contribute a foreword.

The second person to whom I pay tribute is Arnold Broom, of our local history society, a man of bubbling ideas. He has edited several of the Society's annual journals and has helped budding authors to navigate their way through Word to the goal of publication. In my case, his ideas are before you. I would never have got this far without him.

Lastly, I should mention the help we have received from the staff at Orphans Press in Leominster. I have worked with them several times before over the preparation of books for printing for the Bromyard History Society and I have always been impressed with their professionalism.

In this book can be found many pictures of aeroplanes, all taken from my collection of over 4,500 images of British, Commonwealth and U.S. military and naval aircraft from the beginning of Service aviation. I began collecting in 1934. At first, and for at least twenty years, it was centred on the products of "Real Photographs Company" of Liverpool (usually known as RP), a company set up in the early thirties by Capt. Ellison Hawks, a man of boundless energy and of wide technical interests. In addition to postcard-sized photographs of aircraft his company produced photographs of ships and railway locomotives. The war precipitated a flurry of aircraft recognition booklets, many of which attracted my editorial efforts; I found I was acting as an unpaid member of staff – but rewarded with an unfailing supply of free photographs. In the absence of an official library of photos I used RP as an important source of pictures for my lectures throughout the war. In spite of my early trend to write everything down this did not apply to my photograph collection. Consequently, attribution is virtually impossible. Copyright has become a minefield, largely as a result of the worldwide web; by chance I have picked up my own photographs in the most unlikely places all round the world, unattributed of course. To the best of my knowledge all the aircraft depicted are the real thing and neither restored nor replicas; imagine the sight and smell and the excitement of the young aviators. Regardless of this disclaimer, I shall recognise any claim to copyright in a future edition.

My lifelong interest in aviation has naturally been accompanied by a reasonably large collection of books and magazines. I append a list of those which are particularly associated with my story.

Hugh Langrishe, October 2021

Bibliography

Aircraft Carriers of the Royal & Commonwealth Navies, Cdr. D.Hobbs RN; Greenhill Books, 1996
Aircraft Carriers of the World, 1914 to the Present, R.Chesnau; Arms & Armour, 1998

Books published by Air Britain:

The Squadrons and Units of the Fleet Air Arm by Theo Balance, Lee Howard and Ray Sturtivant, 2016
The British Air Commission and Lend Lease, compiled by K J Meekcoms
Fleet Air Arm Aircraft 1939-1945, Ray Sturtivant; 1995
The Squadrons of the Royal Air Force & Commonwealth, 1918-1988, James Halley
Royal Air Force Flying Training and Support Units, Ray Sturtivant, J.Hamlin & James Halley, 1997
Royal Air Force Aircraft, L1000 – N9999, James Halley

Books in the famous series published by Putnam:

The Royal Aircraft Factory, Paul R Hare, 1990
The Aeroplanes of the Royal Flying Corps Military Wing, J.M.Bruce, 1982
British Naval Aircraft since 1912, O.Thetford, 1958/'82
Aircraft of the Royal Air Force since 1918, O.Thetford 1979
Armstrong Whitworth Aircraft since 1913, O.Tapper 1975/'88
Blackburn Aircraft since 1909, A.J.Jackson 1968/'88
De Havilland Aircraft since 1909, 1962/'87
Fairey Aircraft since 1915, H.A.Taylor, 1974/'88
Hawker Aircraft since 1920. Francis K Mason, 1961/'91
Sopwith Aircraftt 1912-1920, H.F.King, 1980
Supermarine Aircraft since 1914, C.F.Andrews & E.B.Morgan, 1981
Vickers Aircraft since 1908, C.F.Andrews & E.B.Morgan, 1969/'88
Westland Aircraft since 1915, Derek N.James, 1991/'95
United States Navy Aircraft since 1911, G.Swanborough & P.M.Bowers, 2nd Edn., 1968/'76

Other Books

British Carrier Aviation, Norman Friedman; Conway Maritime Press 1988
British Military Aircraft Serials 1878-1987, B.Robertson; Midlands Counties Publicns 1987
British Piston Aero-engines & their Aircraft, A.Lumsden; Airlife 1994
Corsair, the F4U in W.W.2, Barrett Tillman; Patrick Stephens 1979
Hellcat, the F6F in W.W.2, Barrett Tillman; Patrick Stephens 1979
Jane's All The World's Aircraft, 1941 Edn., compiled L.Bridgman; Sampson Low
Jane's Fighting Aircraft of World War II; a compilation by Bracken Books 1989
Rolls-Royce Piston Aero Engines, A.A.Rubbra; Rolls-Royce Heritage Trust historical series No. 16, 1990
Royal Navy Escort Carriers, Cdr. D.Hobbs RN; Maritime Books 2003
TBF/TBM Avenger Units of W.W.2, Barrett Tillman; Osprey 1999
The Forgotten Fleet, John Winton; Michael Joseph 1969
The Ironworks, A History of Grumman's Fighting Aeroplanes, T.C.Treadwell; Airlife 1990

The Merlin in Perspective – The Combat Years, Alec Harvey-Bailey; Rolls-Royce Heritage Trust historical series, No. 2, 1987
The Military Airfields of Britain, K. Delve: Crowood Press:-
    Northern England, 2006
    Scotland & Northern Ireland, 2010
    Southern England, 2005
    South-Western England, 2006
The Seafire, D,Brown; Ian Allan 1975
The Times Atlas of the Second World War, Guild Publishing 1989
U.S. Military Aircraft Designations & Serials, J.M.Andrade; Midlands Counties Publications. 1997

Hugh and his brother, John Langrishe, at home after Christmas 1943. John had reached home on Christmas Eve after a journey from Italy where he had been held as a POW since June 1942, having been captured with his RA battery in the Western Desert, N. Africa. At the Armistice with Italy in September 1943 the POW camp was opened and the men had to find their own way home. John chose to walk along the Appenines from the Po valley to the Allied lines on the River Sangro. He eventually got a "lift" in a USAAF transport aircraft from Morocco to Prestwick. John was mentioned in dispatches for his epic walk across the mountains.

Seafire F.III and LF.III of 887 and 894 NAS, HMS Indefatigable after VJ Day in the Pacific

Crash party out of sight...
A Goodyear FGI-D Corsair IV from 1831 Sqn HMS Glory comes to grief in 1945

Index

Aircraft are referenced without their manufacturer's name or their mark numbers, this is to prevent undue emphasis on different manufacturers of the same aircraft.

Military personnel of especial relevance to Hugh's story are included with their most senior wartime rank.

Military Personnel